Wiring for Beginners

Step by Step Guide on How to Wire Your House

(Detail Guide About Home Electrical Installations & Repairs)

Gerda Kreiger

Published By **Jordan Levy**

Gerda Kreiger

Wiring for Beginners: Step by Step Guide on How to Wire Your House (Detail Guide About Home Electrical Installations & Repairs)

ISBN   978-1-998769-05-6

No part of this guidebook shall be reproduced in any form without permission in writing from the publisher except in the case of brief quotations embodied in critical articles or reviews.

Legal & Disclaimer

The information contained in this ebook is not designed to replace or take the place of any form of medicine or professional medical advice. The information in this ebook has been provided for educational & entertainment purposes only.

The information contained in this book has been compiled from sources deemed reliable, and it is accurate to the best of the Author's knowledge; however, the Author cannot guarantee its accuracy and validity and cannot be held liable for any errors or omissions. Changes are periodically made to this book. You must consult your doctor or get professional medical advice before using any of the suggested remedies, techniques, or information in this book.

Upon using the information contained in this book, you agree to hold harmless the Author

Table Of Contents

Introduction

People depend on electricity constantly and, when electricity goes out during an earthquake or there's a tripped breaker or another issue within an electrical circuit, understanding the essential components of the electrical system can help you get things back up and running. It's also crucial to understand who's responsible for which component of your electrical support.

The service company is responsible for the line part of your management and includes all the way is up to the connection point at your home. From there it's referred to by the name of heap and everything located on the heaping side of the line is your responsibility.

Chapter 1: Essential Equipment And Devices

For Wiring In The Home

Electrical equipment is designed to protect your from electric shock, and allow you to work faster and enhance your safety. If you're working on a few electrical projects in your home having a good collection of basic electrical equipment is a smart, sound idea that could help you with your electrical projects and safer

Fundamental Tools
Prepare yourself for everyday household electrical work with the basics: screwdrivers that are protected as well as wire analyzers and pincers.

Screwdrivers with elastic-held handles or that are protected can be found in every electrician's belt of tools. Don't use screwdrivers that have plastic handles , because they may break, posing the risk of a shock. Make sure to use screwdrivers that have electrical protections that have been evaluated for safety against electric shocks that can deliver voltages that exceeds 1,000 Volts.

Wire analyzers tell you that the power has been set for a specific wire . They are vital tools to ensure the health of any electrical

project. There are a variety of analyzers available for use with various purposes.

Voltage analyzers detect the connect wires and link protection in order to demonstrate that wires are active prior to working on them.

Congruity analyzers look at the reliability of switches, breakers and other attachments when power is turned off.

Computerized multi-testers perform similar tests like congruency analyzers and voltage identifiers all in one unit.

GFCI repository analyzers, that are also used to check standard electrical containers and notify you when the power is shut off at the outlet.

Circuit locators inform you of what circuit an switch or outlet can be found located on.

Wire cutters are believed to cut different checks of wire and link.

Wire strippers are similar to wire cutters but they have the advantage of expanding of a middle indent which makes it simpler to strip and cut different types of wire protection, without cutting the wire itself.

Wire strippers and cutters require some time to master, however, once you've mastered them, they'll can make protecting yourself from cutting wire faster.

Forceps are used to hold wires or other objects in place while you work with them or move them.

Side-cutting forceps come with an edge for cutting on one end and are available in short-nosed, long-nosed,
and bent types. Look for a spring within the handle to make the job easier by opening the jaws at the end of each use.

Linesman forceps are large side-cutting pincers with jaws that grasp and also fronts. Linesman forceps can cut wires, and then constrict them. Find a good pairIt should feel somewhat heavy in your hands and you can work without an issue.

Long nose pincers feature large and pointed jaws which are ideal for aiming to get to places and creating small circles of wire which attach the screw terminals.

Be sure to buy two of them to take charge of your household wiring since certain types are designed to accommodate better wires for gadgets.
The fish tape helps in guiding cables through conduits, or running connections through walls much easier.
Extra Tools
In the course of your electrical project It is possible that you need different equipment such as lighting, saws for drywall and utility blade.

The wire-twisting screwdriver, a steel stud on wire-bowing screwdrivers provides a simple and effective technique for twisting, circling, or connecting strong wires when installing switches and outlets.
The cordless screwdriver lets you make electrical connections faster.

Programmable strippers - easily strip coaxial wire in one step using programmed strippers.

Adjustable wrench: A tool with an adjustable head designed to accommodate various sizes of loose pieces.

Drywall Saw : A wood saw that is used to cut holes for drywall installations. The majority of them have an etch point, which means you don't need to bore pilot hole.

Stud locator/laser level − Locates studs and people in walls that have been completed.

A screwdriver with a rotating head Screwdriver that rotates Ideal for the installation of long-strung screws lighting fixtures and switch plates.

Stapler: A stapling device that is designed to accept various lengths of staples to connect electrical connections to joists, studs, or studs.

Link ripper - Removes the sheathing from the link that is non-metallic.

Blade for utility - A blade equipped with disposable cutters that are tradable allows you to easily peel the sheath of an elongated link.

Spotlight - If you are working in an electrical area it is possible to turn off the power completely or work behind a wall, in complete darkness.

Electrical tape Make use of dark tape to guard and create wiring associations. Use the various shades of tape to mark and identify wires. Effective electrical tape is stronger long-lasting, durable and adheres better than less expensive tape.
Pocket for your electrician's device A pocket for instruments will keep your instruments in easy reach.

Safety Tools
Include things such as safety glasses, safety glasses and electrician's gloves nearby Be sure that you use them when working with electricity.

Electrician's gloves are safe and protect your hands from 20,000 to 1,000 voltages, based on the type of gloves you buy.
Eyewear for wellbeing should always be worn when performing electrical work in order to prevent an electrical flash or cut wire from eating or scratching your eyes.

Electrical cables with GFCI security could be an emergency when no GFCI protected outlet is available. A 12-check electrical string is the one you need to make use of to ensure your high-voltage devices have the power they require as well as prolonging their lifespan and keeping safe from fire hazards.

Stepping stool should be non-conductive , such as fiberglass or wood that will aid in protecting you from electric shock. Don't use the stepping stool that is made of aluminum which is an electric conduit when working using electricity.

The elastic soled shoes can help protect your feet from shock.

Chapter 2: The Most Common Terminologies

Utilized In The Field Of Wiring

Electricians typically use specific phrasing to describe the equipment and cycles within their work. Understanding the various terms used by electricians will help you to increase your ability to communicate with other electricians and be aware of how electrical functions operate. If you're working in the field of electrical engineering or are considering becoming an electrician, understanding the different terms used by electricians could help you.

What are terms for electricians?
Electrician terms are terms used by professionals in the electrician industry. They can be used to describe techniques as well as devices or information related to electrical activity. Electric terms allow experts to communicate effectively while working as electricians are usually knowledgeable of the majority of electrical terms.

What is the reason it is important to understand the fundamental electrical terminology?

It is essential to understand the most important electrical terms to understand the language of other electricians and converse with professionals in your discipline. Electricians typically use basic electrical terms.

Working, and knowing the definitions of those terms will make sure you're following proper guidelines, for example using the correct methods and tools.

These are the 16 terms that electricians will use during their work:

1. Volts

Electricians employ volts to represent the force of an electric flow by numbers. They define each volt in terms of the ratio of electrical power that is expected to flow between two conduction marks in which an electric flow flows between the two points of focus. Electricians think of wiring frameworks that are less than 110 volts to be low-voltage, comparable to doorbells. High voltage

Anything that exceeds 1,000 volts, for example, switching on or off a large warming device.

2. Circuit

Circuits are a circular shape of electrical energy. Electricians are able to use electricity when the circuit is running. Common devices that use circuits include light fixtures and cooling units.

A short-out occurs when the circuit isn't an entire circle, like something is hindering the circle from becoming an ideal circle. When people are shocked by electricity, it is a sign that they were an element of the movement of electricity in the circuit.

3. Semiconductor

Semiconductors make up a large part of circuits. They are a solid substance that is able to direct the electricity of a separator to the metal. Gadgets made of semiconductors, which are similar to silicone, are used in circuits that help direct electricity.

4. Energy-saving devices

Electricians make use of energy-saving devices to store and conserve energy in a productive manner, while making use of devices that offer lighting, heating and cooling. Typically, the majority of devices consume electricity even when not in use.

Utilizing energy-saving devices to save money for clients and help machines in longevity, because they're only using a significant amount of energy when using them, and in every situation, they're storing energy.

5. Joule

A joule is a measurement of energy within the International System of Units. It reflects the amount of energy was moved to begin with the first item, and then on to the next one, which is estimated within one millimeter. In this case, for instance, the amount of energy needed to lift an item one millimeter off the ground will be one joule.

6. Watts

Electricians determine the speed of energy movement using Watts. It calculates the amount of joules which can be transformed to power in a consistent manner. For example, a five-watt light could change five joules in electrical power.

Energy is transformed into light power continuously. The more Watts a machine uses the greater amount of energy is generated every second.

7. Wiring

Wiring provides electrical power to machines and switches all through an arrangement. Wiring is the conduit for electrical flow between the ends of an electrical wire and then to the other. There are several types of wiring used by electricians:

Electrical engineers use this wiring using weatherproof links to make short-term events more enjoyable like a party or military installation.

Wiring for the packaging consists of plastic or wood packaging around the wires which makes them sturdy enough to be used in workplaces and homes.
The secure wiring wire requires pins and metal clasps This means that electricians must

use secure wiring when working indoors as it won't be able to handle the harsh weather.

Conductor wire: Electricians utilize steel cylinders to protect wiring and often believe that this is the most secure form of wiring.

8. Circuit breakers

Circuit breakers are devices that safeguard circuits from excessive electric energy. Electricians place circuit breakers through the structure to limit the amount of electricity that flows through the circuit.

Breakers are often the first to decide on the most powerful flow that a device is able to handle.

9. Meter for electricity

Electricians may use electricity meters to determine how much energy a building makes use of. Typically, they show the amount of electrically controlled equipment is in a particular unit. It's like the clock since a structure draws more power. The lines of the meter are moved in the clockwise direction.

10. Conductor

Conductors are the exterior insulation of electrical wires to protect the wires. They are made of a variety of materialsthat are similar to metal or. Electricians take courses to increase their protection from electrical shocks. The courses also protect wiring from regular fraying or weather-related issues.

11. Measure

Tests determine the distance of the wire in order to understand the dimensions of the channel needed for the wire. Typically, the measurements of wiring are 10, 12 or 14. Electricians use measurements to measure examine and width in reverse and when the checking of a wire grows the width will become smaller.

12. Flood defender

Flood defenses are an instrument that provides protection for electrical equipment from voltage spikes. If a machine is damaged, flood defenders are a must.

If it encounters an overly high voltage that is too high, it may cause harm to the device or the power source it is connected to. Connected defenders regulate the voltage and block any abnormally high voltages. Many plug extensions with electrical power sources include flood defenders integrated into them.

13. Switches
They are devices that regulate the amount of energy flowing into outlets and machines. Electricians install switches that be switched both on and off. If an electrician flips on a switch it provides power to the device however , if the switch is not turned on it can disrupt the flow of energy.

For example, a light switch that's switched off will stop the flow of energy.

14. Outlets
Power sources that are electrical, also called fittings and attachments, enable electricity to enter devices by connecting them with an electrical frame. The power source serves as an extension of the device and electricity that is able to supply electricity from an electricity source back to the machine which expects electricity to function.

15. Amp
Amperage, sometimes referred to as amp refers to the term used to describe the

estimate for the total amount of electrons streaming through the circuit, in contrast to the power they're streaming

against. Electricians compose amps by adding the letter A following the number. For example in the event that a vacuum is 12 amps. Electrical engineers may make it 12A.

16. Focus on the load

The heap is the primary source of force that a structure needs to operate from. Each circuit that regulates electricity start from a pile spot. Electricians typically put circuit breakers in the heap community to measure and regulate the amount of power the heap community can transmit.

Chapter 3: House Safety Measures For

Wiring

It is essential to be safe when working using electricity. Security must not be compromised, and certain routine procedures must be followed in the first place. The fundamental rules related to the electrical health and safety that are recorded beneath will be helpful when working with electricity.

1. The first step to ensure electrical safety is to stay away from water at all times while working using electricity. Do not touch or attempt to fix any electrical equipment or circuits using wet hands. This increases the conductivity of electricity flow.
2. Do not use equipment with damaged or frayed cords, damaged protection or broken plugs.

3. If you're slicing away at the contents of your house, make sure to shut off power. It is also advisable to put up a notice on the help

board to ensure that nobody turns the main switch on accidentally.

4. Always use protected devices when working.

5. Electrical risks include exposed components and electrical wiring that is not guarded.

hardware that may be invigorated suddeny. This type of hardware typically displays warning indications such as "Shock Risk". Always be aware of such warning signs and follow the health guidelines set out in the electrical code that is maintained by the country in which you reside.

6. Always wear protective rubber gloves and goggles when working at any branch circuit, or any different electrical system.

7. Never attempt fixing equipment that is powered. Always ensure that the device is not powered by using an analyzer. When the analyzer's electrical circuit comes in contact with hot or live wire, the lamp within the analyzer will be turned off.

The light will illuminate to show that electrical energy is running through the wire. Make sure to examine all wires, as well as the exterior metal covering of the assistance board, and any other wires that draping are using an electronic analyzer prior to continuing in your work.

8. Don't use any steel or aluminum step stool when you are working on a storage facility in your home. A surge of electricity can cause you to be grounded and the entire electrical flow will pass across your entire body. Make use of a bamboo, wooden or a fiberglass stepping stool, all things being identical.

9. Know the wire code of your country.

10. Make sure to check your GFCIs once per month. An GFCI (Ground Fault Circuit Interrupter) is an RCD (Residual Current Device). They are now a common sight in homes of today, particularly in cold areas such as kitchens and bathrooms, because they help keep free of dangers from electric shocks. They are designed to disengage quickly, thereby protecting against any

injuries caused by over-current or short-out issues.

Chapter 4: House Wiring Project

Under cabinet lights
Have you got a space in your kitchen space that could use some real light? Particularly task lighting in an area that requires it, such as in the areas of the ledge?

The first step is arranging. Are you able to find an area under your cupboards that is a valence and is it strong to hide the lighting? What type of lighting you have to consider is contingent on this.

There are many options available to explore. The most up-to-date innovation I

We have seen LED tape which can be cut to specific lengths, and then positioned under cabinets. Bright lighting is also common, however depending on the style and design, it may require a larger area of valence to shield the fixtures in the absence of.

In this design I'm working with 1" valence which is why I searched for equipment made by General Electric that would fit nicely in the space, and are available in 12-" as well as 18" lengths. They use LED lighting technology and can be connected and string-associated to be permanently put up with the purchase of an additional package. They may also be connected to create the desired length of the string

for the cupboard section where you have to install the lighting beneath.

LIGHTING INSIDE THE CUBE LOW-PROFILE LED
Every project is distinct, dependent on the specific circumstances. In this project I will explain the essentials that will guide you through how the project can be completed, as well as the typical issues you could face. Take a look at the video for some of the challenges that you could face depending on your particular situation.
Arranging
The first stage is to figure out how you will get your sources of power for your project, and to plan out the way you'll connect that are urgently sought

The apparatus zones. Under certain conditions it might not be possible without lots of discomfort in the workplace or on the off possibility of not having an acceptable basic configuration, you could have to consider an option for a module or battery-powered lighting.

For this design, I've got an a pair of switches that perform well. The first switch operates on the garburator, which is an entirely different circuit, and the second switch is connected to the lighting circuit which we'll use to power this lighting. The switch is for one light bulb located over the sink area. This light circuit has eight outlets and we've got lots of space for circuit stacking to include these below-bureau

lights. I'm able to swap the case to a three-group case, however, it would involve tons of work that isn't needed. At a cost of about $10, I can purchase a very loaded switch that's two single-shaft switches, incorporated into one. Because there are only six guides within this 2-group box (power into and out to the light bulb and the switch leg drop on the garburator) I have plenty of room to add four guides I'd like to add to handle the new arrangement of LED strip.

Once you've decided on the type of equipment you want to make use of, you need to estimate and organize for the quantity of installations and the lengths at which machines will be required to finish the job. Prepare a list of the materials you'll need and then go

Purchase your supplies from a reliable home improvement retailer. Take into consideration the possibility that your hyperlinks will be protected completely, or perhaps protected, but not able to defend against the actual damage.

Because my feed connections are hidden, but will be visible in specific regions, I selected the reinforced link (BX) The circuit is protected by a 15A breaker, thus the 14/2 flexible aluminum shielded link the one I chose. If you have everything you need, now is the perfect moment to start the process. Be sure to allow plenty in duration (4-8 hours) otherwise you'll cause a mess!

Safety First!

The first step (as typical) is to find out which electrical switch is in charge of the the circuit where air is sucked away and then shuts off the breaker.

Moving things along

Now I'm trying to open the intersection box and double make sure the circuit is not working using an ampli-meter also known as a voltage analyzer. Additionally, I need to close the circuit that is in charge of the garburator in order to keep away from the live wire of that switch as I work inside the box of intersection.

The most memorable part of LED lighting is directly on top of this Crate. As

Each case is a unique piece and unique, but I'm able to easily access this case to get rid of both 12" strip that are going to be placed beneath the bureau. Start by piercing an opening in to the rear of the valence space to an area of wall.

Be aware not to allow the piece to go into the wall more than it is necessary because you do not want to hit any of the links or pipes that could be within the wall cavity.

Then I enter the link entry point inside the crate, which is located underneath the opening. Then I begin to take care of a fishing bandage between these. This isn't easy but, with enough the right amount of tolerance, it will end. Be sure to tape the circular portion of the fish tape closed with electrical tape to ensure it isn't able to find the most recent information about anything within the wall's wall depression, like connections, protection plumbing pipes, other things.

First Hole Drilled, and Fish Tape In
If you're not sure, take the bureau out and cut a hole in the wall that is large enough to allow for your hand should you require. If you do decide to go through this step, ensure that the hole you cut will be concealed by the newly-reintroduced bureau!

After the fish tape is inside place, remove the protective layer off the link, revealing the

guides. Make sure you strip the wires to a sufficient length (8-10") to ensure the wires are ready to connect once they are inside the box for intersection. When using a link that is protected make sure you use the plastic shorts' enemies which are included in the link or are available to purchase when you purchase the link. Check out the video for more point-by-point pts that send highly clad links to intersection boxes.

Link is installed in Junction Box
To make it easier to fish from as well as since I have plenty of space in this crate, I'll begin fishing from the switch box, and then down to the ledge that is behind the dishwasher. At that point the link will pass through the cabinets, behind the sink and then behind another cabinet on the other side of

the sink. After that I'd like to go through the upper cabinets then up to an intersection box which I will introduce above the cabinets. From the intersection bo, x, I will make sure that I have connections down to the lights underneath the two remaining bureau areas.

The establishment of the Fixtures
With each of the links being taken care of and intersection boxes mounted and grafts

created, now is the perfect moment to start introducing the lighting apparatuses. Follow the mounting instructions provided, and make use of layouts for that allow for a slack by the mounting. Make sure to space and measure your installation equally, then package and secure any excess

the wiring of the apparatus and further away, as might be anticipated.

Last Connections
Now I'm trying to create my last connections to the box for switches. I have a 2-wire link, which is the power in, and a 2-wire connection to the light bulb (existing) and I've added two 2-wire connections that connect to my lighting under the cabinet.

I am removing the single shaft switch. I join the four non-partisan (white) cables together. The hot wire connects into one of the standard wires that connect to the hot part of the super load switch. The dark wire to the current pot light is connected into the highest exchange terminal on the opposite side. the dark wire is connected to the top exchange terminal.

The wires that will be used for the new lighting are connected with the ponytail, which connects with the switch's base terminal.

With each of the current associations, we are able to switch to the breaker and test our efforts.

The Results

The addition of the under-cupboard lighting has made a huge improvement in the lighting of this area of the kitchen. It's an enormous project, but the results were well-justified. It's been a great experience in the time since the two houses were built and erected to the point that in the event that you're creating your dream home or tackling major kitchen renovations include

lighting under the cabinet as a part of your design.

How to Install a Ceiling Fan

The Old Fixture is being eliminated
1
Turn off the power by turning off the switch box for electrical appliances. When working

with wiring or electricity it is essential to cut down the capacity of the space prior to contacting anything else. Locate the switch that is enclosing your home and the area or the area where your fan will be placed within.

Some homes may have multiple breaker boxes with one being the primary, and several sub-breaker enclosures situated in better locations. If you have multiple breakers, turn off the power to the sub-breaker box, and at the primary box prior to starting work.

2

Remove before putting on the new apparatus. Use a stepping stool stool that allows you to climb up to the roof so that you can access the installation. Make sure to hold the equipment using one hand while you remove any screws that tie down to the roof.

If they were removed, the previous equipment should have the possibility of securing to the roof.

For the introduction of your brand new roof fan, it is best contact an electrician licensed to install it for you. The wiring should be run through your walls. This is a difficult task and could be extremely dangerous when it gets clogged up.

If you aren't able to hold the device in place while you remove the screws, ask someone else to assist. This can reduce the chance of the device falling off the top and will make it easier to remove the screws.

3

Unplug the wires of the electrical installation that was previously in use. Find out where the wires of the installation are joined to the wires that are coming

on the roof using from the roof using plastic connectors. Make sure the installation is supported with a different method than the wires. Then, start unscrewing and removing every connector made of plastic.

Once removed, remove the device from the roof and then dispose of it. put it in a safe place for a few months without wires, or attempt to upgrade everything on your roof with no assistance from an experienced electrician. If the wiring doesn't look exactly what you expected, you should stop immediately and get an expert for assistance.

Make sure that you have a reliable source of support for any equipment you take out or install while working on them. Make sure you don't

Let the installation be supported with wires, since they could cause harm to them.

4

A roofbox is circular metal fitting that various installations join to. Find any nails or screws that are that are securing the roof-enclosed location and get rid of them. The circuit can be pushed further towards the roof, or try to remove it to remove it.

If you are able to reach the area above the roof and take an excellent look at it the possibility is that it could be less difficult to remove. You can use the spotlight to examine it from beneath and find the most simple method of taking it down.

It is unlikely that any typical roof box can be able to withstand the weight of a rotating roof fan. Don't use the present roof-box unless you're certain beyond a shadow of an doubt that it has been tested as the roof fan.

In the event that you're not sure, take it off and substitute it with another roof box with an rated weight over the weight that your fan is. Contact your local hardware store or search on the internet to locate one that is sufficient.

If you're not sure if you'd like to use your fan, examine it for any marks or model numbers that can aid in

recognizing it. Check the internet or visit your local store for help to make sure you know that your weight ratings of your case is higher than the weight of the fan.

Certain roof boxes contain link braces that can be used to save the wiring arrangement. In the event that you spot a piece of metal tightly securing the wiring, look for screws on the installation's side. Reduce the screw, and slide the link cinch on the top of the wiring to get rid of the wire.

Connecting with New Ceiling Box

1

Purchase a fan box that has holders bar when the device is situated in between two roofs or when it's an

an extendable bar that can be rigidly between two roof joists and will give you an item to connect the fan to. Purchase a fan box that has holders bar at your local electrical shop and make use of it to ensure that your fan is connected between two roof Joists.

Use a small electric light to examine the hole in your roof which is where the new roof box is placed. If you don't find a large piece of timber straight over the opening, it's likely that your roof box is positioned within two joints. Locate the studs on your roof without doubt.

2

Choose a fan box with a screw in the event that you're working with just a single roof joist. If your roof is merely beneath an outlining of wood, you can use the fan box which could be damaged simply inserted in the outlining. Purchase a fan box that does not have the holder bar at the nearby hardware or equipment retailer.

You can see the opening through which you'll introduce your roof box to create an a wood joist that is straight across the opening. If you can easily without any effort get into the joist, use an adjustable fan box with screws.

3

Cut the opening around to make sure it's the proper dimension. A roof box that has been evaluated by a fan could be slightly thicker or more than a normal roof box. The fan box should be held close to the opening and trace it around with the pencil. Make use of a drywall saw for removing the roof overabundantly to ensure that the box is able to be able to fit into.

Take care to not cut through any wires in the roof. Make sure to keep the roof as low as possible to avoid the possibility of accidentally hitting something.

4 Feed the links into an electrical outlet in the box. Then, if you are accidentally trapped, lift the box toward and

Connect the roof. Once you've done that connect the wire to the roof via the hole inside the box on top.

If the fan box comes with the link clasp, you can run the wire through it too. Cut it off from to the point that it rests in the center that is the box and secure the screws to the link cinch to keep the wires in place.

5 Make sure that the roof is secured with the fan in position. Assuming that you have an air conditioner that is connected easily with a roofjoist place the crate on the joist and use the screws provided to fix it. If you'd like to use to use a holder bar, put the bar between two Joists. Make use of a wrench with a programmable setting to rotate the

Bar and then broaden it and pivot to ensure that the bar can be securely by the two joints. Be sure to follow the guidelines of the producer to connect the fan box with the bar that holds the holder.

6 Connect the section or roof plate. The fan's roof plate or section is the device which your fan will hang from. The roof plate should be positioned up in the direction of the fan, and then run the wires in the hole in the middle. Use the screws provided to ensure that the roof plate up.

Always follow the manufacturer's guidelines when working with explicit components. The method of attaching roof plates and sections may change, but you must

Be done with care so that the fan is attached by the roofing.

If you own an emblem on your roof or trim that is designed to cover the fitting, connect it immediately. It will create a better part that is encircles the fitting and connects up to the roof. Use a small amount of urethane-based glue to secure it on the roof, after which you can secure it by using four nails that are complete. For a better finished appearance, seal the nail holes with spackle or caulk.

The Ceiling Fan is collected
1

Purchase a fan kit. It contains all the components you need to get and will also include a roof fan. They should be accessible from

the local store for home improvements or the internet. You should ensure that you purchase an air conditioner that is appropriate for the space in which it will be installed.

Make use of a 36 inch (91 millimeter) fan in a space less than 140 square feet (13.4 square meters)

Choose a 42-inch (110 centimeters) fan for a space that is between 140 square feet (13.4 m2)) and 229 square feet (20.9 m2))

Choose a 52-inch (130 centimeters) fan for anything larger that 225 square feet (20.9 m2)).

2

Download the fan's body. This download is the long , metal line used to keep the fan away from its roof. To keep your fan's body in place and attach the wires to the fan

The download. Set the download up on top of the body. Repair the lock messes on the base of your download, until it's securely connected to the body of the fan.

If you're unsure of the way your fan was made and assembled check out the manufacturer's instructions for more information.

Certain fans could come with different screws for securing to secure the download. Be sure to fix all of them to prevent the download and also the fan body from separating when mounting the fan.

There are several factors that could alter the length you require of your

download. If your roof is calculated to be a certain height, you must use a download to stop the edges that cut off by fans from causing an uproar in the town. If your roof is higher than 8ft (2.4 meters) it is recommended to use a download that's at least 10 inches (25 centimeters) in length to ensure the greatest airflow.

If your roof is that is less than 9 ' (2.7 m) tall, you need consider using a roof fan that is hugger that is typically designed for roofs with lower heights. They will come with a less restricted download, or they may not even have any download at all by any means, which allows them to fit as close to the roof as is reasonably expected.

3

The fan should be lifted towards the roof. Make use of a stepping stool, or step stool to carefully lift the fan that is gathering toward the roof. The majority of fans utilize a suitable method to place them under the roof as you connect the wire. If it doesn't need to be suspended, you can ask a third party to help you set it up in the same manner in this manner.

It could be most comfortable to climb the steps and have someone else to be a fan in front of the side to. Make sure your stool is safe and you are able to stay in it with a friend if you'd like to. If at any time you don't feel you have the right sense of security, contact an electrician to assist.

4

Interconnect the unconnected wires. The unbiased wire provides an exit route to the current coming through the fan. It usually has the appearance of a white covering. Keep the two wires that are unbiased in good condition and connect the closures with striped stripes. Use a wire connector made of plastic to ensure they remain in place and secure it using electrical tape.

Wiring plans can change with different roof fans. Check out the help provided by the manufacturer in case your wires clash with one another, or contact an authorized person if you want to join with the ground wiring. The ground wire is likely to be green or completely exposed and is used to prevent electric shocks. Find the ground wires, tie them around, and then fix them with an

Connectors for wires made of plastic. Use electrical tape to hold the wires as well as the connector in place.
Make sure that the ground wire that is coming from the roof has been connected with the screw for ground inside the fan box, since this connection is crucial in order for ground wiring to meet the gap. In the event that there isn't a ground screw, or don't know the best way to properly ground a wire contact an electrician who is licensed.

6

Connect the wires that are not connected. The extra wires are hot wires, which are utilized for capacity purposes to the fan as well as any fittings for lighting that are connected to it. They will generally be dark, but they could be a variation in different wiring schemes. Constrict the excess

Connect wires and secure them by using a wire connector made of plastic or electrical tape.

This technique will switch control for the fan as well as any fixtures that form part of it. Get the help of your manufacturer for advice on wiring different control options or seek advice from an electrician.

7

Connect the wires to the box for your fan. Carefully push the wires from the roof to the inside of the case to collect them. While doing this, make sure the electrical tape are secure and there's no wire exposed.

Uncovered wiring can cause dangerous brief stays, as well as additional problems. If you observe

43

If you see any wires exposed, swiftly fix it and completely cover it.

The Mounting of the Fan

1

Install the shade on the box that houses your fan. The shade will rise off the download and then cover the wiring and the attachment that is used to support the fan's configuration. Use the screws provided to attach your overhang with the fan box.

2

Join the edges of the fan. Separately lift the edges of the fan to their mounts on the engine. Set them up as recommended by the manufacturer's guide and then use the screws to attach them to the engine securely.

Check that the screws are as tight as you can get as close to. Screws that are loose will cause the edges of the fan shake and could even come loose when the fan is operating.

3

Introduce any lighting installation. There will be space to install a light on the bottom. They will for the most part be incredibly simple to connect and put in place. Consult the manufacturer's guide for instructions on the most efficient way to connect the lighting system into the fan you have chosen.

The wiring of the lighting device is generally adheres to the same guidelines for wiring the fan. Join the wires of the

Similar to each other and connect them using electrical tape and connectors to avoid short-circuiting.

4

Take a walk and test your fan. Return to the electrical sweatshirt-circuiting and reestablish power in the important segments of your home. Then, turn the fan on to the wall and observe it for a few seconds to verify that it is working as it should.

In relation to rooftop fans, in the event that already have one however, you don't have control over the cut, pulling chains, or using switch on the walls, how do you go is the best way to switch a current fan completely to a remote-controlled model, or a different one? The establishment of a roof remote controlled light and fan?

Instructions for Installing A Ceiling Fan with Remote Control

Safety First!
The first step (as typical) is to determine which electrical switch is in charge of the power of the circuit that we are working on and then shutting off the breaker.

The breaker is shut, and the chip-away segway board has been locked (assuming this is indeed the case you have it) this means that it's impossible for anyone to be a victim the circuit while you're working on the circuit, you are able to safely begin working.

In the event that you do not have plans for locking off access for the board place an insulating tape over the handle of the breaker and stand solidly on it in an off-hand situation and put up a notice informing others to keep the breakers off.

ELIMINATE the COVER WITH THE FAN
The next step is to remove the shade on the fan, so that you can see the wiring in place and to create a plan for the controller's beneficiary unit. For this particular fan, an edging ring hides the screws for mounting that keep the shade cover up.

Make the ring turn a 1/4 turn until it is free , and then it will remove the screws. All you need to do is be at the end of one screw.

either side, and then reduce the two sides that pass through the key-opening or L-spaces on the shade cover so that it to come out and expose the mounting section as well as the fan's wiring.

Eliminate the overhang of the roof fan
Remove the wiring that is in place
Determine how the fan will be connected with the transmitter. There is a 3-wire connection within the box for power sources. This provides a standard non-partisan (white) wire. Additionally, the fan's light source is linked to the red wire, while the fan is linked to that dark transmitter.

We only need one of these new hot wires to function as the official power source for the unit that will be receiving it. I'll use the dark wire

(and the uninvolved and the transparent) in order to regulate the collector. The additional wire (red) is covered by a wire nut and will not be employed in this location. This sketch shows how the fan was connected to the switches initially.

The remote receiver is connected to the remote
Now we disconnect the light pack and fan and prepare to install the collector that is far away. The white wire of the link to stockpile will connect with line neutral and that dark one will connect directly with live in or inline connection from Merom to create the hue.

The independent output of the collector is currently in conjunction with the standard

The colors of the light and fan and that dark wire for the fan will be connected to the wire labeled "fan" (red) The blue wire for the light will be connected with the wire that is named after the light.

With each of the associations formed in the process, we'll need to wrap the wires one further into the container, and place the distant collector within the space available within the mount section. It can be a difficult job to get everything fit, but it is essential that everything is arranged in a pleasant way and that all wires are not trapped or squeezed too tightly to keep them away from harm to transmitters, potentially creating a short-out situation.

Remote for the Roof Fan Remote in the correct place
When the collector that is far off is in place and the trim and shade ring are back in the right place We can introduce our enhancements inside the box for the multiple-posse switches.

In this particular setting the two switches that control the light and fan are located inside a three-posse box, which is also equipped with a switch that controls the containers that are exchanged in the room. Remove the faceplate as well as the device screws, and make sure that the switch that controls the light has an orange wire on one side, while the switch that controls the fan has an unlit wire on the other side.

Each switch will be connected by the dark hot wire of the same hot join that is associated both switches. We must eliminate the switch that we'll never will need, and also cap the wires by securing them with wire nuts. We will leave the red wire in the same manner just like we did in the outlet for fans.

The remaining switch to act as a professional control for the collector that is far away. I will shift the switch we're as but not using to the middle position, and save this switch to be used for repository in the original position, then use a clean filler plate to take the space left by the switch we have eliminated.

With the switch once more at the correct position and the faceplate now in place it's the ideal moment to take a look at our company.

Re-establish the circuit's capacity and then switch on the specialist switch in the wall. The controller should now regulate the activities of the fan as well as the light. The speed and direction of the fan must be set to the optimal position using switches for draw and converse and the pull-chain for the light unit must be turned on for the remote to function.

Installing a wireless Doorbell

1

It is a straightforward way to find the space that is the location of the doorbell switch. Doorbell switches are the switch you push to activate the doorbell. Choose a prominent location close to the doorway to the switch. The guests should

You can choose to detect it as they are staying in front of your entrance.
Making the doorbell move up eye-level with the opposite side of the door jamb can earn you a good payoff.

It is recommended to select an outdoor doorbell which isn't harmed by snow or rain.

2

Connect the switch by using screw or glue. The switches on the majority of doorbells are accompanied by openings on the back, making the installation simple. Find the openings for the switch and use the drill with an electrical motor to install the switch either in one direction or to the wall. In contrast apply the large areas of

the strength of a paste on the back of the switch. Then, apply it firmly to the desired surface.

Clean the area you're joining the change with a spotless, soiled substance prior to the installation.

3

Find a place to display the rings box. In the ideal scenario the ring box needs be placed near the focal point of your house to ensure that everyone hears it. Select a space that's usually the same in terms of separation from the numerous rooms in your house.

Select a location in which you aren't typically closing the door to ensure the sound can be heard.

You could, for instance, create a toll area around your lounge area or lounge.
Make sure you check the reach on the ringer to confirm it's ringing the doorbell.

4
Place batteries into the ring box and hang them on the ring box and then mount. Most remote toolboxes accept D batteries. Then, open the box and place the batteries in according to the instructions and then close the backboard in a safe manner. Find a place in your house where you think the sound is coming from and then join the container using screws.

Installing a wired doorbell

1
Disconnect the power source from the wire box or breaker to prevent injuries. Make sure that the circuits that are responsible for supplying electric power to any source that

you're working with are shut off prior to starting the installation. Turn off the appropriate switches on your breaker board and wire box.

Check the light switches or other outlets in the vicinity to make sure that electrical power has been turned off.

2

Connect the doorbell wires to rings. Remove the cover from the rings and then connect the wires via the aide channel until they reach the appropriate terminals. Fold the edges of the

The wires are placed over the terminals. Install the screws holding the wires into the correct position.

There are a variety of entryway tolls , with a variety of options and sounds.

Some rings contain tiny wiring graphs inside for ease of installation.

The cover must be pulled off the rings easily without the need for instruments.

To be able to refer back later, mark the wires with a pencil by writing the areas where you want everyone to be (for example , transformer or doorbell switch) with tiny pieces of cover tape, which are joined to all of us.

3

Place the rings in their proper place. Make sure you have the wire connected to your transformer to pay your tolls. Place the new rings in the desired location to join them , and use the screws supplied to join this gadget onto the roof or wall. After the toll plate is removed, put the cover onto the gadget and gently push it until it is fitted properly.

4

Connect the doorbell switch to your entrance. Select a spot for the doorbell switch, near your entranceway. Create an opening for the wires that emerge from the back of the switch to the wall. This is close to the transformer and rings. The majority of models will

are screws that help to fix the plate in place. Introduce the screws using an electrical drill and then place the cover over the device until it snaps in place.

5

Connect the wires to ensure that the transformer is connected to both the ringer and the doorbell. Carefully fold the ends of the wires to the transformer's terminals. This tiny gadget can alter the AC power that is coming from the entrance to lower voltage to regulate the bells.

Transformers are typically mounted easily onto electrical boxes to keep high-voltage wires enclosed.

6

Connect the tolls and switch by using bend wire connectors. Use bend wire connectors made of plastic to connect wires between rings and switch without difficulty. Connect the ends of the wires and put the cap over the closures, and bend to ensure that the wires remain securely connected.

The immediate connection will create the signal between the doorbell and the rings, and the transformer will impede this link to make it the voltage that is protected.

7

Re-establish power and test the doorbell. Reestablish the power by using your power source.

Breaker or wire box. The doorbell should be pushed to test the structure. If the capability of the ring is adequate, the task is completed. If the doorbell isn't working then turn off the power and then test the wire association.

Installation of an electrical socket outlet
Get ready for a secure Installation

1
Review the guidelines for restrooms and kitchens. Because of the potential for water-related spills, these places require extra security precautions. The same rules are recommended for spaces with no air circulation, open-air

areas that include sheds, pantries and many other places near hot t, sinks and other sources of water.
In the minimum, you'll need an outlet that has an GFCI (ground shortcoming circuit interrupter) also known as an RCD (leftover

present gadget). It shuts off power in the event that it is wet.

installing high-quality attachments in these areas should be done by electricians who are certified. Replacing damaged attachments in these areas could be feasible without the help of any other person.

2

Be safe from electric shocks. Forestall electrocution by avoiding the risk of:
Make use of devices that have handles made of elastic.
Wear shoes with elastic soles.
Avoid contact with the skin that is not covered with metallic or other conductive surfaces even multimeter tests.

3

Turn off the power source. Turn off the electrical switch, or remove the wire driving the source of power you'll cut away. If you're not 100 % certain of the source you should eliminate, turn off the power source for your entire house and use an electronic lamp.

4

Check the voltage. Don't assume that the wires are not working without conducting tests. First, test a live circuit to ensure that the device is operating, and follow up with a examine the circuit you're testing.

taking care of. If you notice an electric current the power source, it is not dead and cannot be addressed.

Non-contact Voltage Analyzer isn't difficult to use, but it is not as reliable. When you're grounded, you can place the device in each of the openings within the source of power. If it is lit, or again, assuming that the display is something other than zero it means that the source of power is active.

The multimeter can be more reliable and provides a more precise result. To test voltage using the multimeter set it at it's AC voltage setting within the 100V region. Check the test by setting the test in red to Live Attachment (the small).

upward opening upward opening US attachment) Then, keep it in place while you install the test in dark first in the attachment that is non-partisan (the larger vertical opening) and, after this point, you can put it into the opening on the ground (rounder opening).

Warning Note: The UK and some of the previous UK states, a small number of houses are wired on the rings circuit. These DIY tests aren't sufficient in these instances. Do not work on circuits in this manner before an electrician spotted the type of circuit.

5

Remove that old connection. Once you're sure that there is no power take off the faceplate of the old attachment, and then pull it off of the wallbox. To disconnect your wires, remove the terminals just a bit.

enough to allow you to slip the wire circles off of enough to let you slip the wire circle off of.

The wiring of the New Socket

1

Be aware of the live, nonpartisan as well as ground connections of the connection. A common current source intended used for home use should include three terminals to connect with the appropriate wires.

US attachments

Metal terminals are active (hot)

Silver terminals are impartial

The green terminals are ground

UK attachments:

"L" demonstrates life

"N" demonstrates unbiased

"E" (or three) equally lines, demonstrates earth (ground)

2

Modify your plan if you have more terminals. If you are able to see a larger quantity of terminals than shown above, you could be to be in one of the following situations:

When replacing a present connection in UK it is common to install two wires of the same type to connect the terminals. The installation of a second attachment will only require only one set of wires.

An US two-attachment outlet for majority of them has a metal tab that connects both live terminals and another one for those two neutrals. If there's a gap, there's a one wire of a particular kind on your wall could connect this wire to the other terminal to manage the two connections.

The GFCI (RSD) outlet is equipped with two terminal arrangements. Line terminals are used for these directions. They are also known as heap terminals. (generally separate with yellow tape) can be used to connect various devices with GFCI assurance.

Take off the closures on your wires. In the event that wires are frayed or scratched, take the damage off, then remove approximately 3/8 " (2cm) in the protection. This can be done with the wire stripper or utility blade. Make sure not scratch the metallic of the wire. This could result in electrical discharge.

imperfections problems later. Make sure you are not under-stripping, so you can fix it later. A few outlets can be used as an aid to place the wire inside the short scoring on the back and mark the end on the row that aids to create your strip's edge. This guide could be intended for an "push-in" connector instead of the recommended wrapping technique.

If you assume that the three wires are wrapped in a PVC coat, locate the finished of the the grounding wire made of copper. Make this happen using long nose pincers. Pull down until you can in the crease of coat to expose the various wires.

Twist the wire until it is in the position that of an umbrella's handle. The most efficient method to secure your wires to be tucked in is to wrap them around the screw.

terminals. To make them ready make sure you turn the stripped end in the U shape, so that it can fit snugly over the screw.

Wire strippers are able to open up within them due to this. Then, slide the finish of the wire into and then turn. If you don't own an electric stripper, you can use long-nose pincers.

Many outlets have push-in connectors or small holes beneath the terminals that secure the wire using spring braces. In the event that you decide to use this feature option, simply push the wires into the holes. In addition, these clips may be weakened and eventually weaken the connection.

The wires should be positioned over the screws in a clockwise direction. Each wire should be placed snugly on its terminal with all three U-twist sides being in close proximity. Wrap them around the direction that screws fix (typically clockwise) to ensure the best contact with the screws. Before doing this ensure that you are around 700% certain that you're using the correct wire:

USA:

This live link appears dim (assuming you have two active links and the second one is red)

Unbiased link that is dark or white

Ground link is not insulated. green or yellow.

Chapter 5: Some Fundamental Concepts

To give you a better idea of where you should start, let's look from the beginning! In this chapter, we'll go over some of the most basic notions regarding wiring and the way that the electrical system in your home functions. Here are a few of the most basic ideas.

The Electricity Industry: How Do They Work?

At its most basic stage is simply a flow of electrons that move from one location to the next. It doesn't matter if we are aware of that or not, electrical energy is all around us all the time. It is true that electricity wasn't developed in the 18th century by Thomas Edison, electricity is the most fundamental element of our environment. It's not just about the impressive displays of lightning that illuminate the sky at night. there's an electrical spark within each of us, in its form in the shape of own neurons, which use electrons to transmit or receive data. These electrically driven sparks of data running through us in all time.

The the air that we breath is filled with electricity. Electricity is always there as a force in nature but it's only recently in the scale of everything that humans have figured out how to harness this force to our own benefit. Electricity is not a mystery anymore to us. Much like how our forefathers were able to master the use of fire, we have been able to master the power of electricity. This understanding has resulted in many advances but it is still a need to be improved. As you read this book, you will discover how to refine this power of nature.

Understanding Your Electricity Grid

The next step to understand electricity is to understand how your home uses it. In the majority of homes you are connected to an electric grid that is controlled by the utility service provider in your area. The utility company frequently delivers power through the grid to your home via the "service access point". We'll discuss the service entrance more in-depth later in this book, but for now, it is important to be aware this:"service entrance" is a simple term "service entrance" simply means the underground wires that serve as the main entry to your home's electric.

The wires go through the utility meter in your home, which is a monitor of the amount of power your household uses. The electricity from here is routed into your "service panel" which is then divided into various "circuits" throughout your house. These circuits of electricity are through which we connect in our devices as well as run our appliances.

The home you live in should be equipped with a central "circuit breaker" which is able to turn off each circuit within the home, at any point. When we are further along in the wiring installation process, you'll be instructed numerous times to check your circuit breaker and shut all your electrical circuits. For now, let's concentrate on how everything functions.

The Art of Gauging Your Wiring

Wires are available in a variety of dimensions and shapes depending on the purpose they are intended for. The gauge of gauge is what you measure its length as well as the amount of electricity it can safely transport. Certain wires are more dense than others. Thin wires that are small in size are generally used for purposes like doorbells, and other smaller-scale installations that don't require a large amount of power. Wires are also available with various kinds of covers.

In general, it is the non-metallic sheathed type of wiring that is frequently utilized in homes. Another kind of wiring that can be employed is commonly referred to as "armored wire" desire is more difficult to handle, but it offers higher security. The wire is able to be stepped on or stomped upon and endure the elements without any issue at all. Make sure you have this sturdy piece of wiring immediately.

Making use of Conduit Systems

Conduit systems are installed to shield the wiring exposed or exposed in garages, basements and attics. The most commonly used type that is used for conduits are "electrical tubing made of metallic". This type of conduit is sold in 10-foot pieces length, which are able to be linked with special attachments known as "couplings". The conduits serve dual purposes of protecting insulation and also allowing for proper conductivity for electrical flow.

The most significant drawback with conduits is that they're rigid and difficult to bend. To bend the conduit, you require a tool known as"bender" "bender". This tool, appropriately named, will bend the conduit wire in the direction you require it to travel.

The tools you will require
There are many tools of trade that are used in wiring, but to keep things simple to the point of being overwhelmed, we'll start with the most fundamental. It is possible to make adjustments and modifications to these recommendations as you progress. These are the most basic tools you'll require.
Power Drill
The majority of people are at least acquainted with the power drill. These drills are speedy and easy to operate. They are equipped with a choice of tip sizes that allows you to create a variety of types of holes in wood.
In its most basic form it operates through an electric motor inside that rotates the drill bit around at a fast speed to force through wood. In addition to drilling holes it can be equipped with special attachments, like the

"screwdriver" edge that is placed so that it is able to quickly adjust screws and make them looser.

The drill is turned on by pressing"on" or press the "on" button in the hand of the gadget. If you own an electric drill that plugs in to the wall's receiving coils directly via the socket. If you're using an electric drill, the power source is a battery that is placed into its handle. Make sure you own your own power drill to complete the projects described in this book. They can be found at the majority of hardware stores and department shops.

Fish Tape

Fish tape is an essential tool used to direct and guide new wiring. If you are wiring your home, the fish tape is likely to be the one that you'll use the most that any other. This device is easy to wrap and is ready for use as soon as you receive it. The tape is a very thin and long steel wire that is neatly wrapped around a wheel.

The reason this tape is referred to as "fish" is due to the fact that in the course of using it, you'll have to hunt for your correct route as you pass the tape along your conduit. This is an essential bladesmithing tool, don't run homes without this!

Pipe Cutter
Pipe cutters are, like their name suggests, are utilized to cut pipes. They can be used to create clear cuts and are considered as the most efficient device to employ in working with pipes during exercises in home improvements. When it comes to wiring, you'll use the tool couple of times. Along with holding pipe in place the instruments are precisely controlled tools in the field of wiring. Keep this in your mind.
Basic Tools

In just one area the most essential tools you'll require to have on hand, I'd like to include: screws driver and hammer as well as tape measures. These are all common household tools and always have a place for any home improvement project. Make sure you have at least a small toolbox of these essential tools on hand. It's really not difficult more than a few essential items and you've got yourself a basic toolbox!

Electrical Tester

The ability to examine whether you are testing AC as well as DC currents, either one or the other electrical testers are employed to test amps as well as the polarity and force that electrical electrical circuits exert. They are essential in finding out which wires are in use and the amount of power flowing through them.

Be sure to are carrying all the items listed in this shopping list in your shopping list, in case you don't have them already. Since when you're installing or wiring of your home, this basic tool can make your life much simpler. You won't need to guess when you've got an electrical tester. You can be exact. It's an excellent thing!

Understanding Your Service Entrance

As we mentioned earlier in this book this is the place at which your home gets electricity (or "service" from your electric utility that is located in your area. In this chapter, we will look at the function of your service entrance bit more in-depth.

Find out if you have An Aerial Service or a Buried

Most of the time the majority of the time, buried service cables are more reliable than aerial service cables due to their being located underground. They are able to provide better protection from weather than aerial cables that suspended over "through through the air" over lengthy, hanging pieces of wire. However however, underground service cables are likely to be more expensive than aerial service cables which is why if you're with a tight budget you're probably due to this reason that aerial cables might appear as a better choice for you.

Service cables that are buried are routed through an underground pipe, and connected directly to the grid of your utility company and then upwards and out to an adjacent transformer. Being able to hide your cables underground will not only keep your from having an unattractive eyesore of cables that hang over the top they also keep them secure from weather that can knock them over or animals posing a risk of a bite! It's true that there are dead squirrels around who were involved in exactly this kind of activity and ate the exposed power cables.

"Overhead" or "aerial" cables are actually air-borne and are surrounded by wooden poles that are hefty and connect the wires from your home and back to the electric company's grid. These wires that are exposed are, of course, susceptible to all sorts of weather conditions, from rain, wind and light, as well as squirrels who decide to make an enticing meal of the wires. If it's not clear prior to beginning your wiring project , make sure that you know whether you've got underground or aerial service.

Calculate Your Amps , and choose the Cable You Want to Use

Before starting, you have to determine your amps and select the kind of cable you will use based on that. Certain cables can handle more amps, while others can handle less. If you want to install wiring for a dryer and washer for instance, you're likely require some high-end cables to support the highest level of amps.

If you're connecting standard circuits to your living space, the cables must reflect that. Texas Instruments have a great calculator that can do the computations for you. The calculator is sold in most department stores, and can also be purchased from a variety of sellers on the internet.

Make sure you take care of your Meter Base

The Meter Base is the rectangular box made of metal that is attached to the sides of houses that are residential. Many people are shocked to learn that it is true that the Meter Base is not owned by the electric corporation, they are owned and operated by the property owner and is the home owner's responsibility to care for it.

Maybe a reason to not purchase a home? If you own the property of the property, you will need to be aware of the base of your meter. You can have nothing more. If you look after your meter base , it takes good care of you!

Be aware of panels and Subpanels

When wiring your first home, you might be tempted by the overwhelming task when you have to deal with all the subpanels and panels which comprise the electrical system of your home. This chapter will guide you through the steps.

How big is your box?

When installing a brand new electrical box, you have to remember that the size of the box has to be in line with the amount of wiring required. If you're dealing with a large amount of wiring, in order to stay safe and make sure you have enough space to work in then you must ensure that your box is large enough.

These aspects must be thought of in advance to ensure that you're equipped to deal with certain of the more complex and challenging parts of wiring. Before you start anything else, you must make sure that you have the proper size box!

Select the Panel that is the best for you.

At the end of the day, when all the fundamental basics are covered, you can choose the right panel for you boils to your personal preferences. It is possible to pick an extremely small panel or a larger one. Choose the one that is best for you.

You're the one who will have to bear it in the end So it could be something you enjoy! There are a lot of options to pick from.

Mount Your Main Panel in Place

The installation of the main panel on your house will be requiring you to put it outside of your home either on its own or with the service meters. The panel serves as the receiver of three big electrical wires, and also serves as the route for the small cables that extend onto the underpanels. Other cords are routed into the meters mount to ensure stability of the entire apparatus. This way, everything comes together without any issues.

Connecting a Breaker Circuit

The circuit for breaker is the component of your switch's electrical circuit that prevents the flow of electricity or "breaks" the electric current in the case of overloaded circuits. This is a crucial component of every electrical system, and in absence of it you expose yourself to the possibility of overloaded circuits , and possibly the possibility of electrical fires.

For the first step, switch your main switch to the "off" setting. This switch is located in the uppermost part of your circuit breakers. Once you have established that your circuit breaker is safely shut off, you can take off the cover of the circuit box for the breaker.

Take your "electrical tester" (as described previously in chapter) and test the voltage, which is an additional security measure to be sure that no electricity is flowing through. Then, put an extension cable through the hole for the conductor, and then secure it.

You'll need to pull out an "lock-nut" to complete this job be careful. Then, you can run the clamp to the hole. Tighten your lock nut. After that, you can run four conductor cables through the Sub-panel Breaker Box.

Know Your Subpanel

Subpanels are usually used to extend their reach over the primary panel. The heavier duty 200-amp panels are able to handle more power. Therefore, it is important to understand the number of amps you can expect to get for the new subpanel. This way, you'll have a think ahead and try to avoid any interruptions. It's safer instead of being sorry.

Connnecting the Service Panel

This section will let me to be transparent. Connecting the service panel isn't an easy job. This is the one step that many feel intimidated by and then end up throwing away the towel and calling the help of a professional electrician.

However, even though it's difficult, it's not impossible. We'll make it through it. So, let's get started! To begin, you'll need to switch on your main breaker and remove the main fuse. A word of caution this time: even after pulling the main fuse, the wires on the top of your box is still in use and will disable it.

Don't, I repeat, do never ever reach out and contact those live wires! Okay, now that you have that set move to the new wiring and strip them as cleanly as you can. Also, make sure they're of the correct length to reach their connections. Simply remove"knockout" cable "knockout" cable that is on the back of the box and then secure it to the box.

After that, you'll need to set up your fuse box set up. This will require connecting your black cable to the screw-terminal point of the fuse holder. Then, begin to screw the white wire into the neutral bar, along with all others white wires. Then connect your main grounded cable and you're all set to go.

The Process of Getting Grounded

If you find an outlet with three points of insertion It is grounded. If you look at the cable for your laptop, you're viewing a grounded connector. Electronic components have been grounded due to an reason. It was in order to protect the environment and ensure safe electrical usage in mind that these items were created. In this chapter, we will look at this feature along with other vital security features in more detail.

Be aware of your grounding System

An organized system can be a useful system. Let me explain it for you. It is evident that in this arrangement, grounded wires in each circuit connect back into the grounding station that is neutrally within the panel for service. This is then located in the ground , so that it's actually grounded.

The service panel is comprised of a long braided cable placed beneath the soil. This kind of grounding system utilizes seven-foot buried rods composed from copper. Most often, only the top of the rod can be seen. Make sure you know the operation of your grounding system in order to properly keep it in good working order.

Make sure you are practicing safety Standards What ever. Safety is always the top priority. Make sure all equipment and cords are disconnected prior to starting. This means removing any live wire connections that remain. If you don't have it done so, turn off your circuit breakers and then manually disconnect all fuse circuits in order to keep yourself safe.

It may be beneficial to check your wires a few times prior to testing to ensure that no power is able from coursing through them. It's worth repeating that you should never use live wires! If, for any reason, you require an increase in the level of service that your panel is receiving, you'll require contacting an expert. However, whatever your decision is, ensure that you follow safe standards.

Ground this, but don't ground that!

Based on the NEC (National Electrical Code) these are systems that need grounded, however there are certain systems that do not require. There are certain situations where grounding is prohibited and it is an excellent idea to be aware of the many aspects.

Alternating Current systems with less than 50 Volts, for instance, should be grounded when used to be used as aerial conduits. They must also be grounded when they are used as transformers within an un-ground system. Any wire that is premise-wide over 150 volts needs to be wired accordingly.

On the other side, if you're designing a grounding system to be used for industrial purposes, such as melts, tempering smelting, and similar processes, you might not be required to connect your electrical equipment after all. In the case of an industrial textile plant, grounding would be strictly forbidden due to the danger of combustible fibers! Keep in mind that the rules don't necessarily have to be set in stone. Take the time to think about the things that need to be grounded as well as what doesn't require a grounding!

Step-by-Step Guide to Wiring

Alright! Now that we've gone over the basics about the essentials of wiring. Let's get started and start the actual wiring task itself!

Here is your completestep-by instruction on wiring. Let's get started!

Locate Your Outlet Box

At first glance the idea of finding an outlet might seem straightforward however, this isn't always the scenario. There are a few situations where homes have been renovated where the outlet was secured, in this case the case, you might be faced with a lengthy look to locate the outlet box.

In most cases, the outlet box is located in the middle of your house. This is the central element of your electrical system. Locate it and make a note of its location to help you start. It's as easy as it gets.

Installing Your Switches

The first thing to complete when installing switchesis to must identify the power source and the fixture it will direct the power to. It is necessary to create an "loop" within the current by connecting you white cable to source and the black wire will connect to the fixture as well as the switch.

In the circuit, you must have the neutral wire connected to one side, as well as the live wire on the opposite side. Once you've done this, put the wires in the switch, and then close the screw onto the circuit's face plate. Your switches are now set up.

Configure Your Circuit Breaker

To set up the circuit breaker, you'll first have to shut the power source to your grid. You must go to the circuit breaker's main panel, and turn all switches off. Then, take a close look at the layout of your circuit breaker and examine the areas for spots which are not being utilized. Take a close look at the lower part and also the cover. Pay attention to any "knockout" plates which could be on the cover.

Then, remove the cover of the electrical panel. Then, take out an electric tester and check the breaker to see if there is any residual power. Select the correct mounting point for the breaker and switch off the handle of the breaker. After that, you can set up your circuit breaker in an appropriate alignment with the other face on the wall. Once you've completed this, turn off the breaker and then replace the cover. The configuration of the circuit breaker is completed.

Install You Wiring

To begin start by turning off the power breaker. Then, test the circuit with a tester to be sure that the current is off. After you have established this then you can pull the wiring. For the majority of projects around the home, you'll require standard 14-gauge wire.

When installing your wiring, make sure you follow your "color code" and connect your wires to the counterparts, which have a color coded connection point. As was discussed in the previous paragraph, using"the "black wire" and the "white wire" as an example the wires will be matched to kind.

Black and red are likely to have active cables and white ones are grounded cords. Yellow blue and yellow are used to make switches, as discussed in the earlier section. Once you have your wiring in place you can then connect it through the outlet.

Also, turn off the power source to your breaker, remove the circuit's faceplates and connect your wires through the circuit. That's the procedure for installing wiring. It's not really that difficult it's just a matter of having to be able to navigate it, and I'm hoping for the best results.

Chapter 6: Electrical Wiring Systems And

Electrical Wiring Methods

Electrical wiring systems are generally standardized, with a range of regulations, norms, and regulations in effect. In accordance with the electrical codes and standards, electrical wiring has to be carried out in a safe manner and with a guidance. If wiring for electrical purposes is completed improperly or not in accordance with any standard, it can cause electrical shocks or short circuits damage to the device or appliance, or malfunctions, which all can reduce the life of your device.

Before the actual installation procedure for commercial, residential, and industrial wires, various factors must be considered. These are the types of ceiling, building, floor, and wall construction, wiring procedure and requirements for installation, in addition to other things.

Let's look at the basics of electrical wiring including the concept that electrical wires are important, as well as the various stages involved, the techniques employed, as well as the most commonly used forms that electrical wires are used in.

NOTE: This is not an electrical wiring user's guide or a lesson. It is simply a theoretical article explaining the various Electrical Wiring Systems and methods to install Electrical Wiring. If you're considering a task that requires AC Mains Electrical Wiring, it is highly recommended to seek expert guidance.

Electrical safety safety tips

Before starting any installation process the foremost consideration should be employee security. Electricity can be harmful and getting in contact with it even when the power is switched on could cause severe injuries, as well as death. To ensure that your workplace is secure take the steps listed below.

1. Always wear protective equipment like goggles, gloves and footwear to avoid direct contact with active or current-producing circuits.

2. Recognize the live components that are exposed in electrical equipment by using techniques and techniques.

3. When connecting wires or installing them make sure to turn off the source of power.

4. The principal switchboard, which must be equipped with circuit breakers, needs to control the flow of electricity by the systems.

5. Conductive materials and tools should be kept away from the live circuits or equipment components.

6. Make sure to use hand tools that are not conductive that are designed for electrical use. If they are used with or current (or or current) ratings that are not stipulated, the tool's electrical insulation could fail and cause electric shock.

Electricity Distribution

The Electricity Board / Department supplies electric power to consumers' property's exterior (either residential) commercial or industrial). The consumer has to connect at that point to their home's main distribution system or switchboard.

Many kinds of electrical load like lighting, fans room coolers, refrigerators connect to the main switchboard or distribution board using cables and wires.

Different kinds of wirings can be used for connecting various electrical devices to the mains and they are suitable in both industrial and residential electrical wiring. Some of them are covered.

Wiring System Electrical Wiring System

Electrical wiring is an integral part of any structure, regardless of whether it's a residence (individual housing units or houses) or a massive commercial construction (office buildings) and industrial structures (factories). Electrical wiring is utilized to power lighting and other circuits in a variety ways and in a variety of systems.

The kind of wiring that is used can have an important influence on the total project's cost. This is why it's crucial to understand which kinds of Electrical Wiring Systems are appropriate for the task at hand.

Below are some of the most typical elements to take into consideration when choosing an electrical wiring method:

* The cost for the wire system
* Types of cable and wire employed
* Wires of high-quality
* The type of burden (light and HVAC, motors etc.)
* Wiring System's Safety
* Future modifications and extensions are also possible.
* The duration of the installation

* The building's construction wood bricks, concrete, mortar, etc.
* Fire prevention
It is essential that the Electrical Wiring System should be capable of defending against wear and tear in normal operating conditions regardless of the type of wire and wiring used.

Electrical Wiring Systems are usually defined by the kind of wire utilized (or at least their class). Industrial, commercial, residential and auditorium wiring systems comprise the following:
* Wiring of Cleats
• Wire Casing, Capping, and Capping
* Battens for Wiring (CTS and TRS)
* Concealed conduit wiring (Surface or concealed)
* Wiring using the lead sheath
Let's examine each of these Wiring Systems or Installations in their own.

Wiring Cleats
Wood, porcelain, or plastic cleats can be attached to ceilings or walls in intervals of 0.6 millimeters. Each cleat is used to support and hold the wire, permitting PVC wires to pass through the holes.

This is a cost-effective wiring method that is utilized to install temporary installation. It isn't suitable for household electrical wiring, and it is outdated.

Capping and Wiring Casing

The cables are routed through grooves within the wooden casing. The wooden casing is designed in a manner that it has a length of a certain length, and parallel grooves that allow cables to run through. Screws attach the wooden casing to ceiling or walls.

A cap made of wood with grooves is positioned over the cables after the cables have been inserted into the grooves of the casing. This is also a cheap wiring method, but short circuits are a major chance of fire.

Wiring Battens

Insulated wires run through those straight wooden battens used in this arrangement. Screws and plugs connect the wooden battens to the walls or ceilings. Link clips made of brass are used to connect the battens' cables.

Nails that resist rust are employed to fix the battens to the clips. If you compare it with other electrical wiring methods, this wiring installation is easy and cost-effective. It takes much less time to put in. They are mostly employed in indoor environments.

Cabtyre Sheathed Wire (CTS) or Tough Rubber Sheathed Wire (TRS) is typically employed as an conductor for electricity in this type of wiring.

Connecting Conduit

PVC wires are connected to PVC pipe conduits or pipes for steel within this wiring. Conduit wiring on the surface or underground conduit wiring are two possible conduit wiring options.

Surface conduit wiring is conduit pipes that are positioned across the ceilings and walls. Concealed conduit wiring happens when conduits are placed beneath the surface of ceilings and walls and are covered with plaster.

In industrial settings Surface conduit wiring is utilized for connecting powerful motors. Concealed wiring on the contrary is the most popular and extensively used method of wiring homes. Conduit wire is considered to be the most secure and most attractive method for wiring (concealed conduit wiring).

Sheathed Lead Wiring

In addition to the kind of cable or wire the wiring design is similar to CTS and TRS wiring.

The electrical conductor is wrapped in Vulcanized Indian rubber before wrapping it in a sheath of Lead-Aluminum alloy (95 percent Lead and 5 percent Aluminum).

Like the Batten Wiring is run over batten made of hardwood and is secured using clippings that are tinned.

Different Types of Drawing

Electrical drawings are helpful in electrical installation projects since they provide the best way to connect different devices and equipment to mains. Drawings show the entire outline of the electrical installation, and assist in the assembly of the various pieces of equipment.

Below are a few examples of wiring schematics for electrical use. Before you begin to learn about these diagrams, you should first know and comprehend the various symbols that are used to create of drawings and the wire connections. Explore a range of wiring symbols for electrical wiring.

Diagram of Blocks

It is a practical drawing that illustrates and describes the devices or equipment's principal working principles. It is composed by blocks which represent primary elements or functions, and are joined with lines that show the connection between the blocks.

The diagram is usually created prior to a circuit diagram being used. It doesn't provide specific details about the system , nor will it provide information on the minor components. This is why the majority of technicians are not interested in this diagram.

Schematic of Circuit Diagram

The following is a simple visual illustration of an electric circuit. It includes dimensions (in millimeters, centimeters and millimeters) for various components like lights, receptacles, light fixtures the junction box, ceiling fan and so on.

Diagram with lines

It's also referred to as a One Line Diagram or Single Line Diagram. it's an easy diagram of the electrical system. It's like an exploded block diagram, except that the typical schematic symbols can be used to represent a variety of electrical gadgets like transformers switches lights and circuit breakers, fans and motors.

It's made up of symbols representing the elements and lines which represent the conductors or wires which connect them.

A block diagram can be used generate lines in the diagram. It doesn't include an outline of the part or wiring details for the components.

However, you can make the wiring according to the steps in this diagram. Diagrams like these are usually used to show the way in which an electric circuit operates.

Schematic of wiring

A diagram of electrical wires is an image representation of the circuit, which shows how the components, elements and equipment are linked.

It gives detailed information about wiring, so that users can easily comprehend how to connect devices. It includes information about the device's relative placement, arrangement, and terminals of devices.

It shows power supply and earth connections, as well as control and signals (with simplified forms) unutilized contact or lead terminal, connection via blocks, plugs, sockets lead-throughs, terminal posts and more.

Schedule Wiring

It's a complete list of the wires and cables that are that are used in the installation as well as their numbers of reference lengths, types and the amount of insulation stripping that is required to solder. It also identifies the raceways for the wire and the beginning and end locations.

The connection of equipment (such as heaters and motors) along with finishing and starting point of reference is illustrated in a wiring table on various sophisticated equipment. It also includes the wire labels, color and sizes, in addition to other things.

List of Components

Although it's not an illustration however, the list of components is a crucial element of the drawing because it defines the many components and symbols that are utilized in other drawings like lines diagrams and wiring schematics as well as block diagrams.

It lists the different types of circuit components and their references numbers. This list is used to identify or locate the actual component that is as well as to show it in other electrical designs to ensure that the proper components are identified prior to starting with the wiring.

The Wiring Preparation

The process of preparing wires, cables and electrical equipment is the next stage in the process of electrical wiring when we look at the steps involved in wiring, including understanding security, understanding the various the various types of wiring systems and understanding the distinctions between different electrical symbols and designs.

The following points are considered in the wiring process.

1. A single conductor made of solid wire or one that is stranded can be utilized (which is comprised of several stands). The single solid wire is strong and are utilized in areas such as power switch contractors where strong connections are needed. For electrical installations the stranded conductors commonly utilized.

2. The specifications of wires will be decided by a range of parameters, such as how many strands within the conductor, the type of insulation the wire's cross-section area the strand's diameter and so on.

3. Pick wires that match the color codes defined by different standards, like blue for phase, brown for neutral green for earth and others. For more information on the electrical wiring colors of cables or wires, click here.

4. The majority of the electrical equipment is essential for installing work such as strippers, cutters and testers, pliers, and more. We have previously written articles for more details on these tools for electricians click here.

5. Find out the dimensions and ratings of electric boxes, switches and Receptacles.

Utilize the wire diagrams before connecting the parts. Install the installation once you have selected the components, tools and cables, all the while making sure to consider and follow the safety guidelines for personnel and equipment.

Certain types of electrical wiring

We are aware that electrical circuits are an enclosed line through which electricity flows through the hot or phase wire to the apparatus or device and then back to the source by the neutral wire.

Fixtures like switches, receptacles and switches junction boxes and other electrical parts can be found on the roads. This means that wiring can pass through these components prior to connecting with the device or equipment.

Based on the way in which devices are connected to or powered by the source, the wiring can be classified into two kinds. They include:

Parallel wiring

• Wiring in series

Many devices on this installation can be powered through a single circuit, which is parallel wiring. It is the most frequently used wiring system in business and homes as equipment is connected in parallel to the power source as shown by the figure.

Receptacles, fixtures, and other devices are connected to those electrical boxes (junction boxes) from where the phase (or hot) and neutral wires are run.

A Series Wiring system is an often utilized wiring technique in which the hot wire goes through various devices before being connected to the neutral wiring at the final device's terminal. It's similar to old lighting for Christmas or serial light wiring, in which one light's failure causes the whole network to go out of commission.

Electrical Wiring Examples of Wiring

Here are a few instances of wire circuits frequently utilized in our homes and offices to get a better understanding concepts of wires.

One-way switches control one light bulb (or the other loads).

As shown in the diagram In the diagram, the hot wire is connected with one of the switches. The other terminal is linked to the bulb's positive terminal, while the bulb's negative terminal is linked to neutral wire.

A single switch is used to control two Blubs.

As shown in the diagram as shown in the diagram, two bulbs connect in parallel to two supply cables (phase as well as neutrals) that are then routed via a single switch.

Two-way switches regulate one Blub.

Staircase wiring is a different name for the wiring. Two two-way switches are utilized to control the light bulb or lamp from two distinct places or sources. The wire is utilized in bedrooms to switch on and off lamps coming from two sources (at the bedside and on the switchboard). The lamp's connection to switches can be seen below.

The wiring of a warehouse

The large godowns, long corridors or warehouses, as well as tunnel-like structures that have several rooms or sections make use of this type of wiring. The lights are switched between one side and another in a sequential sequence.

When someone leaves one room to enter another room, the light switch in the room before is shut off, and the lights in the new room are on. The lamp in one room is turned off and then turns on the other. The wiring diagram for the warehouse is illustrated in the diagram below.

A One-Way Control Switch controls a Fluorescent Lamp

In the image below, a lamp can be switched with a one-way switch, ballast and an capacitor. The switch's one end is connected to the wire for phase while the other connects to the choke (or ballast). In the diagram the lamp electrode is attached to the choke and one electrode is connected for the neutral wire.

The wiring of a socket outlet

When power is delivered to the socket through the switch, the outlet holds a plug , and it is able to send the current. Single sockets and sockets with radial connections can be shown in the figure below.

Connecting a Control Board

The following image shows an illustration of the controller switchboard. A ceiling fan as well as a fluorescent lamp and an electric light bulb are controlled by switches in this set-up.

Chapter 7: A Brief History Of Home Electrical

Wiring

Electrical service to American homes began in the late 1890s and blossomed from 1920 to 1935, by which time 70 percent of American homes were connected to the electrical utility grid. In the following 200 some years, the methods for installing wiring in those homes has seen several important innovations aimed at improving the safety of electrical systems.

Knob-And-Tube Wiring
Between 1890 and 1910, a wiring system known as knob-and-tube was the principal system of installation. It was ?uite a dependable system for the time, and a surprising number of American homes still have knob-and-tube wiring functioning, where it is often found alongside more modern updates.
In knob-and-tube wiring, individually conducting wires protected by rubberized cloth fabric are installed in stud and joist cavities, held in place by porcelain knob insulators attached to the sides of framing

members, and protected by porcelain tube insulators where the wires run through framing members. In this wiring system, hot wires and neutral wires were run separately for safety. The system also allowed long circuit runs to be constructed by splicing together lengths of wire. To do this, the insulation was stripped back, a new wire was wrapped around the exposed bare wire, and the splice was soldered together then taped to cover the splice. The downfall was the wire was exposed and there was no ground wire used.

Where knob-and-tube wiring is still functioning, it is living on borrowed time, since the rubberized cloth insulation used on the wires has an expected lifespan of about 25 years before it begins to crack and break down. Electrical systems containing functioning knob-and-tube wiring are in critical need of an upgrade. But just because you see knob-and-tubes in some wall or floor cavities, doesn't necessarily mean you are in danger. It was common practice to simply leave old wiring in place when a home was rewired. It's possible that the porcelain insulators and wires you see are merely antique remnants of earlier wiring installation. An electrician can tell you for sure.

Flexible Armored Cable (Greenfield)

In the 1920s to 1940s, electrical installations took a turn to a more protective wiring scheme—flexible armored cable. Flex, also known as Greenfield, was a welcomed addition to home wiring because the flexible metal walls helped to protect the wires from damage, and also offered a metal pathway that could ground the system when properly installed. Although it was an improvement, this wiring method had its troubles. Although the individual wire conductors are protected, the flexible outer metal jacket serves as a proper ground only when the metal pathway is complete all the way to the service entrance and grounding rod. There is still no separate ground wire in these installations.

First-Generation Sheathed Cable

In the 1930s, a ꓺuicker installation method was developed. Nonmetallic-sheathed cable was born, which incorporated a rubberized fabric coating sheath, much like knob and tube wiring, but here the hot and neutral wire were run together in this one sheathing. It also had its drawbacks due to the lack of a ground wire, but its development would eventually lead to major innovation. Early sheathed cable, however, also has an expected lifespan of about 25 years, and where it is still in use, such installations need to be upgraded.

Metal Conduit

The 1940s brought the age of metal conduit. This invention allowed users to pull many individual conducting wires in the same rigid metal tube enclosure. The conduit itself is considered a viable grounding method, and the system can also allow another separate grounding wire (usually an insulated green wire) to be pulled through the conduit. Conduit has been in use ever since those days and is still the recommended method for wiring in certain applications, such as when wiring needs to be run along the face of basement masonry walls or in exposed locations. Most homes have some areas where conduit is used, though it is now sometimes made with rigid plastic PVC conduit rather than metal.

Modern NM Cable

The newest addition to wiring was introduced in around 1965. The form of NM cable was an update to older NM cable, incorporating the use of a bare copper grounding wire that joined the insulated hot and neutral wires contained within the sheathing. Instead of rubberized sheathing, modern NM cable uses a very tough and durable vinyl sheathing. This update made the MN cable inexpensive and very easy to install. It is a very flexible product and is used extensively in virtually every new home built.

Along with NM cable for interior use, a related type of cable was also developed for underground use. Underground feeder wire (UF) can be buried directly under the ground without the need for a protecting conduit. This type of wire has a hot, a neutral, and a ground wire embedded in a solid plastic vinyl sheath that protects it from moisture. This offers an inexpensive method for running power underground to outbuildings and yard lights.

Metals Used in Wires
Through most of the history of residential electrical service, the preferred metal used in the conducting wires has been copper, known

as the best conductor of electrical current. In the mid-1960s, when copper prices were quite high, aluminum came into vogue as a material for electrical wiring. Residential installations between 1965 and 1974 sometimes used wires that were solid aluminum, or aluminum covered with a thin layer of copper. Aluminum (AU) or copper-coated aluminum (AL-CU) wiring is perfectly safe if connected to receptacles, switches, and other devices rated for use with aluminum, but it can pose problems when it's installed with devices intended for use with copper wiring only. Because of these issues, aluminum or copper-clad aluminum is no longer used in residential applications. If you have aluminum wiring, repairs are best made by a professional.

Modern Innovations
Copper wire conductors in NM sheathed cable or in rigid metal or PVC plastic conduit has been the norm since the mid-1970s, and there are currently no new innovations in the wiring materials themselves. Recent safety improvements have involved the extended application of GFCI (ground-fault circuit interrupter) devices, and more recently, AFCI

(arc-fault circuit interrupter) devices that help protect against fire and shock by sensing changes in current flow and shutting off power before problems occur.

But the history of residential wiring is one of the periodic innovations that can revolutionize the industry. It is possible that another such innovation looms on the near horizon.

Electrical Wiring Types, Sizes, and Installation
Much of what you need to know for electrical repairs and remodeling involves wiring—how to identify it, how to buy it, and how to install it with proper connections. If you're planning any electrical project, learning the basics of wiring materials and installation is the best place to start. Understanding basic wiring terminology and identifying the most common types of wire and cable will help when investigating wiring problems and when choosing the wiring for new installation and remodeling projects.

Here are all the basic elements you need to understand about electrical wiring.

Understanding Wire Sizing
The proper wire size is critical to any electrical wire installation. Wire sizing indicates the

diameter of the metal conductor of the wire and is based on the American Wire Gauge (AWG) system. The gauge of a wire relates to the wire's current-carrying capacity, or how much amperage the wire can safely handle. When choosing the right size of the wire, you must consider the gauge of the wire, the wire capacity, and what the wire will be used for.

Wires that are not properly matched to the amperage of the circuits they serve can create a notable risk of short circuit and fire.

Non-Metallic (NM) Sheathed Cable

Most interior wiring is done with non-metallic, or NM, cable—also known by the popular brand name "Romex." NM cable is made of three or more wires wrapped inside a flexible plastic jacket, or sheathing. It is used for most interior circuits, such as those for outlets, switches, light fixtures, and appliances. Learn the basics of NM cable to choose the right type for your next electrical project.

Electrical Wire Color Coding

Color coding is used both on the outer sheathing of bundled electrical cables and on the individual conduction wires within cables or inside conduit. Understanding this color coding can help you identify what the wiring is used for and helps maintain consistency within an electrical system.

Cable coloring relates to the size of the wires inside the cable and the cable's amperage rating. For example, white-sheathed NM cable is used for 15-amp circuits, while yellow NM cable is rated for 20-amp circuits.

The coloring on individual conducting wires usually does not indicate a size or rating but rather the standard or preferred use of the wire. For example, black and red wires typically are used for current-carrying or "hot" connections, and white wires usually are grounded "neutral" conductors. Green-insulated wires and bare copper wires are used for grounding wires.

Understanding Electrical Wiring Labeling

Electrical wires and cable have markings stamped or printed on their insulation or outer sheathing. These markings provide important information about the wiring and insulation, including the wire size and material, the type of insulation, the number of wires contained (inside a cable), and any special ratings or characteristics of the wire.

While looking at the color of wire or cable will help you narrow down the options at the store, reading and understanding the labels on wiring is the best way to ensure you get the properly rated material for your project.

Direct Burial Cable

Standard electrical cable is designed to be run indoors, where it stays dry and is protected by wall, ceiling, or floor structures. For outdoor projects or when running wiring underground, you must use direct burial cable, which can be installed underground with or without conduit (depending on local building code rules). With direct burial cable, the individual conducting wires are embedded in solid vinyl to fully protect them from moisture.

How to Strip Electrical Wire

Stripping electrical wire involves removing the plastic insulation surrounding the wire's metal core. It's important to do this carefully so there is no damage to the metal. The procedure is simple but requires a special wire stripping tool and an understanding of how to use it. This is a critical skill—and tool— for DIYers to have for any wiring project.

Maximum Number of Wires Allowed in Conduit

When running individual electrical wires inside conduit, there is a limit to how many wires are allowed. The maximum allowable number is known as the "fill capacity," and this depends on several factors, including the size of the conduit, the gauge of the wires, and the conduit material. Metal (EMT), plastic (PVC), and flexible conduit all have different fill capacities, even when they're nominally the same size.

Wiring an Electrical Circuit Breaker Panel
The electrical panel, or service panel, is the power distribution point of a home electrical system. This is where all of the individual circuits of the house get their power and where they are protected by breakers or fuses. Wiring an electrical panel is a job for a licensed electrician, but DIYers should have a basic understanding of how a panel works and the critical role that breakers play in any system.

Electrical Disconnect Switches
An electrical disconnect switch provides a means to shut off the power to a home's electrical system from an outdoor location. It is typically mounted below the electric meter, either on the side of a home or on the utility company's power pole. Not all homes have a dedicated disconnect. They are commonly used when the service panel (which also

serves as the main disconnect) is located indoors and therefore is not accessible to emergency responders or utility workers. Like electrical service panels, a disconnect must be installed by a licensed electrician.

Understanding Electrical Wire Labeling
Wiring sold for electrical projects often carries labeling to help you choose the right product for your needs. Letters, numbers, and wording on wiring labels tell you important information, such as the wire material, the size of the wire, and the type of insulation used on the conducting wires. Labels are found on both individual insulated wires and on insulated cable containing bundles of wires. Cables carry labels indicating the cable type or construction as well as the number of wires inside the cable.

Labels on Non-Metallic Cable (Romex)
The most common type of wiring used in homes is non-metallic (NM) cable, commonly called "Romex," after the popular brand name. New NM cable contains two or more insulated conducting wires and usually a bare ground wire. The wires may be wrapped in paper, and all of the wires are encased in a flexible plastic jacket or sheathing.

The labels on the outer sheathing of NM cable indicate the size, or gauge, of the individual conducting wires, the wire material, the number of wires contained inside the cable, the maximum voltage rating, and whether there is a ground wire present. The wire size and number of wires are indicated with numbers. A ground wire is indicated by "G," "w/G," or "with Ground." The wire material is indicated by "CU" for copper and "AL" for aluminum.

Here are some examples of labels on common cable types used in home wiring:

• 14-2G: Cable contains two insulated wires plus a ground wire; the wires are 14-gauge.

• 14-3G: Cable contains three insulated wires plus a ground wire; the wires are 14-gauge.

• 12-2 w/G: Cable contains two insulated wires plus a ground wire; the wires are 12-gauge.

• 12-3 w/G: Cable contains three insulated wires plus a ground wire; the wires are 12-gauge.

• 600 V: Cable is rated for a maximum of 600 volts; this is standard for residential NM cable.

• TYPE NM-B: Non-metallic type-B cable; this is the current standard for residential installations. "NM-B" cable is more heat-resistant than older "NM" cable.

Underground Feeder Cable

Most NM cable is used in "dry," or interior, locations, where the cable is protected inside wall, ceiling, and floor cavities. Underground feeder (UF) cable is a special type of non-metallic cable that is suitable for "wet" locations, or for unprotected locations like direct burial in the ground. UF cable is usually gray (not white, yellow, orange, or black, like standard NM cable); it is labeled "UF-B" and may include "Sunlight Resistant" or similar wording. UF cable uses the same symbols as standard NM cable to indicate the number and gauge of wires.

Labels on Individual Wires

Individual insulated wires are used in home wiring when an installation calls for conduit— a rigid or flexible protective pipe or tubing through which the wires are run. Electricians buy the individual conducting wire by the spool so they can pull different wires from different spools as needed.

The important labeling on individual wires relates to the wire insulation—the plastic coating that covers the metal conducting wire. The most common types of wire used in home wiring include:

- THHN
- THWN
- THW
- XHHN

Here's what the letters on the labels mean:

- T: Thermoplastic insulation, a fire-resistant material
- H: Heat-resistant; able to withstand temperatures up to 167 F.
- HH: Highly heat-resistant; able to withstand temperatures up to 194 F.
- W: "Wet," or approved for damp and wet locations; this wire is also suitable for dry locations
- X: Insulation made of a synthetic polymer that is flame-retardant
- N: Nylon-coated for resistance to oil and gasoline

Labels on Low-Voltage and Thermostat Wires

Low-voltage wiring used around the home includes small non-metallic cable used for thermostats and other control devices and paired insulated wire used for landscape lighting systems. Wire for landscape lights usually is black and has labeling stamped into the wire insulation. Labels typically include:

- Wire size: Indicated by a number (such as 12, for 12-gauge) or a number followed by "AWG," for American Wire Gauge.
- Number of wires: Usually indicated by the number 2; landscape wiring typically has two insulated wires stuck together (similar to a lamp cord) and contains no ground wire.
- Properties: Wording indicating sunlight-resistance or suitability for underground installation.

Thermostat cable is similar to NM cable but contains four or more small insulated wires and no ground wire. The cable may or may not be labeled. Each wire has its own color to help you connect to the appropriate terminal at the thermostat and the equipment it controls. Although color coding is not universal, the lettering on the thermostat terminals is relatively standard:

- C: Common wire; allows for continuous power flow from the R wire; not all thermostats use this terminal
- R: 24-volt power supply from the furnace transformer
- Rc: Calls for heat or cooling; there may be more than one Rc terminal
- G: Fan
- W: Heat

• Y: Air conditioner

Electrical Wiring Color Coding System
Opening up an outlet or light switch box, you might be confronted with a bewildering array of wires of different colors. Black, white, bare copper, and other colors closely intermingle, yet each one has a specific purpose. Knowing the purpose of each wire will keep you safe and your house's electrical system in top working order.
Electrical Cable and Wire Color Markings
Non-metallic (or NM) 120-volt and 240-volt electrical cable come in two main parts: the outer plastic sheathing (or jacket) and the inner, color-coded wires. The sheathing binds the inner wires together, and its outer markings indicate the number of wires and size of wire (gauge) within the sheathing. The color of the sheathing indicates recommended usages. For example, white sheathing means that the inner wires are 14-gauge and yellow sheathing indicates that they are 12-gauge.
But looking deeper, the color of the wires inside of the sheathing reveals that different colored wires serve different purposes. The National Electrical Code (NEC)says that white

or gray must be used for neutral conductors and that bare copper or green wires must be used as ground wires. Beyond that are general, industry-accepted rules about wire color that indicate their purpose.

Black Wires: Hot

Black insulation is always used for hot wires and is common in most standard household circuits.

The term "hot" is used for source wires that carry power from the electric service panel to a destination, such as a light or an outlet. Even though you are permitted to use a white wire as a hot wire by marking it with electrical tape, the opposite is not recommended or allowed. In other words, do not use a black wire as a neutral or ground wire, or for any purpose other than for carrying live electrical loads.

Red Wires: Hot

Red wires are used to designate hot wires.

Red wires are sometimes used as the second hot wire in 240-volt installations. Another useful application for red wires is to interconnect hardwired smoke detectors so that if one alarm is triggered all of the others go off simultaneously.

White Wires With Black or Red Tape: Hot

When a white wire is augmented with a red or black color marking, this often indicates that it is being used as a hot wire rather than a neutral wire. Typically, this is indicated with a band of black or red electrical tape (but other colors may be used) wrapped around the wire's insulation.

For instance, a white wire in a two-wire cable may be used for the second hot wire on a 240-volt appliance or outlet circuit. This white wire should be looped several times around with black electrical tape to show that it is being used for something other than a neutral.

Bare Copper Wires: Ground

Bare copper wires are the most common type of wire used for grounding.

All electrical devices must be grounded. In the event of a fault, grounding provides a safe pathway for electricity to travel. The current passes back to the ground or earth. Bare copper wires connect to electrical devices, such as switches, outlets, and fixtures, as well as metal appliance frames or housings. Metal electrical boxes also need ground connection because they are made of a conductive material. Plastic boxes are nonconductive and do not need to be grounded.

Green Wires: Ground

Green insulated wires are sometimes used for grounding.

Ground screws on electrical devices are often painted green, too. Never use a green wire for any purpose other than for grounding.

White or Gray Wires: Neutral

White or gray indicates a neutral wire.

When examining a white or gray wire, make certain that it has not been wrapped in electrical tape. This would indicate a hot wire. Older wires sometimes may lose their electrical tape wrapping. So, if the box has a loose loop of tape inside of it, there is the possibility that it may have come off of the neutral wire.

The term neutral can be dangerously deceiving as it appears to imply a non-electrified wire. It is important to note that neutral wires may also be carrying power and can shock you. While wires designated as hot (black or red insulated wires) carry power from the service panel (breaker box) to the device, neutral wires carry power back to the service panel. Thus, both hot and neutral wires have the potential to shock and injure you.

Blue and Yellow Wires

Blue and yellow wires are sometimes used as hot wires inside an electrical conduit.

Rarely are blue and yellow wires found in NM cable. Blue wires are commonly used for travelers in a three-way and four-way switch applications.

How an Electrical System Works

Everyone uses electricity in their homes every day, but how does it get there and how is it distributed throughout the home? For electricity to function properly, it must always complete a circuit.

Electricity flows in from one of two 120-volt wires and backs out through a grounded neutral wire. Any flaw in the wire to and from these points will interrupt the current's path and cause a fault in one of your circuits.

Knowing how the power flows into your home, how it's connected, and how it is distributed can help you isolate any problems that occur.

Service Entrance

The utility company's overhead service lines feed the transformer to step down the voltage to feed your home. It then travels to the weather head (service head) which is attached to a conduit connected to a meter box. This assembly is attached with anchor bolts and straps to support the weight of the pipe and wire.

Two 120-volt wires and a grounded neutral wire feed the meter through the weather head. The utility company is responsible for power to the meter, and the homeowner takes it from there.

The service from the utility company to the meter is always live unless it comes and turns it off. If there appears to be a problem on their side of the meter, don't hesitate to call the company to repair the problem. It has special eＱuipment for just such repairs. Never attempt to work on their side of the meter, ever!

Electric Meter

The electric meter is attached to the service entrance pipe and is usually to the side of your house. It may be attached to the utility company's power pole also. It can be fed overhead or underground.

The meter is a watt measuring device supplied by the utility company to track each month's power consumption. There are meters with numbered dials such as a watch on older models and new state-of-the-art digital meters that can be read right from the utility company's office.

Weatherproof Disconnect

In most cases, the utility company will require a weatherproof disconnect right after the meter connection. This is often referred to as a safety switch or service disconnect. This allows the homeowner to disconnect the power from the utility company from the outside of the house without having to get to the electrical panel.

A great reason for this would be a house fire. The fire department can kill the power from outside the home without entering the home. This allows them to spray water on fire without worrying about being electrocuted.

Electrical Panel

Known as the electrical panel, breaker box, fuse box, or service panel, this piece of equipment is the next device in line. This panel's job is to distribute power throughout your home and disconnect power from the incoming feed.

The power comes into the main breaker and is usually 100 or 200 amps. Individual breakers then distribute individual circuits (called branch circuits) throughout your home.

These breakers range in size from 15 to 100 amps. Lighting circuits would be 15 amps, outlet circuits would be 20 amps, and a sub-panel circuit to a garage or tool shed would usually be 60 or 100 amps.

Grounding Wire and Water Ground Connection

The service must be connected to a ground rod outside the house and also bonded around the water meter in the house. A jumper connected on both sides of the meter must be made to allow the meter to be removed without losing a ground connection.

Approved Electrical Boxes

The branch circuits are run into electrical boxes that are mounted inside of the walls of every room of your house. The National Electric Code reꟼuires that wires be spliced into boxes.

The reason is to make every connection accessible. For instance, if you splice the wires together and tape them within the wall cavities with no box and cover it with drywall, how will you get back to it to work on the splice if there's a problem? You can open a box at any time.

Switches

Switches come in many different styles. There are single-pole, three-way, four-way, dimmer, and motion-sensing switches. Their purpose is to turn on and off a circuit from different places in your home. Switches are used to control lighting, ceiling fans, receptacles, and appliances. Switches have different amperage ratings depending on the load requirements.

Receptacles

Receptacles, commonly referred to as outlets, are used to provide individual plug-in points for power distribution. The housing market most frequently uses 125-volt as well as 15- and 20-amp receptacles for general household equipment. For appliances such as 250-volt window air-conditioning units, a 250-volt 30-amp outlet is required.

Hopefully, having learned the basic parts of the electrical system will be useful to you in the future. Knowing how everything flows from start to finish helps in tracking down electrical problems that might arise.

Home Electrical Wiring & Connections

It's time to tackle some wiring projects in your home, but where do you begin? It's important that you know what you're dealing with before you start, so a little lesson in 'Wiring

101' is in order. From understanding the different types of wires you'll find to installing switches, outlets, and a few major appliances, let's look at the basics of home wiring.

The Common Wires in Your Home

Before you begin your first DIY electrical project, you should learn a little about the wires you'll be working with. Wires vary greatly and each is designed for a purpose.

The wiring in your home is chosen to accommodate the load it must carry as well as the conditions it will be exposed to. Some are designed for indoor use while others can be buried. Some are for your panel while others hook up your lights and outlets.

It may be confusing at first, but you will probably deal with only a few types of wire in your home.

Understand Colors and Labeling

Electrical wire has very convenient ways of telling you what it is. Most of the coding is standard, so with a little study, you'll be able to figure out what you have to work with.

Wiring does not come in a variety of colors to make it look good. No, there is a wire color coding system that applies to most wires in your home. Most importantly, you need to know that the black, red, blue, and yellow wires are hot and green is often the ground.

If you look, you will also find a series of letters on a wire. These labels are also standard and will tell you more information about the makeup of the wire. For instance, the code may tell you whether it's aluminum or copper or whether or not it is heat resistant.

As you learn more about wiring, you'll realize just how often you need to know these things.

Wire Size Matters

It is critical in any wiring project that you match the gauge of the wire with the amperage rating of the circuit. Failing to do so can lead to a fire.

A wire's gauge is the physical size of the wire, but the scale is opposite of the wire's circumference. This means that a 2-gauge wire is actually larger than a 14-gauge wire. The size determines how much current can pass through, so the larger wires will be used for your heavier loads.

Installing an Outlet

Many homeowners want to take care of the basic wiring needs for their house. Among the most common projects is installing an outlet. It's a basic project that almost anyone can do if they take the time to understand the process.

You may also want to understand how to wire a split outlet. This comes in handy if, for instance, you want to plug a lamp into an outlet and be able to turn it on from a wall switch.

Simple Installation of a Single-Pole Switch

Light switches are the other electrical installations you might want to handle yourself. The majority of these in homes are what is known as a 'single-pole switch.' They're just as easy to replace as an outlet.

Somewhere in your home, you may come across an odd looking switch that makes you stop and wonder. It's likely that this is a three-way switch. They're a little more complicated and used when multiple switches control a single light.

When You Need to Install an Electrical Panel

The majority of homeowners will not mess around with the electric meter or service disconnect and leave these up to the utility company or hire an electrician. However, you might work with the electrical panel.

Whether you're installing a new panel or making repairs on an old one, it's important that you get it right. After all, this is the hub for your entire home's electricity.

Be sure to properly label any connections you make and update them with any changes. Accidentally turning off the lights on your wife in the bathroom when you meant to disconnect the kitchen may lead to some choice words.

Installing a Dishwasher

New appliances come with their own electrical challenges, which is why many people choose to pay for the installation. If you're a true DIY-er, you can install a dishwasher without problems.

The dishwasher comes with two hook-up challenges: the wiring and the water and drain lines. That's why it's a good idea to choose a location near your sink. It will save you time and money.

DIY Wiring for Your Oven

Your electric range may also require your electric prowess. You might find yourself replacing the oven bake element, which is a relatively simple project.

You may also need to connect the cord. The exact method you need to follow is going to depend on whether you have a 3- or 4-prong cord.

Keep in mind that these large appliances carry heavy voltages, so read up on the safety tips. For instance, plugging a loose cord into a receptacle to check the fit can give you a deadly shock.

Don't do it.

Installing Dryer Cords

Why oh why do you have to buy a dryer cord separately? It's one of the great mysteries of home improvement, but it's a fact of life. Next time you need to install that new dryer, you'll be prepared after this tutorial.

Home Electrical Basics

People depend on electricity constantly, and when the power goes out in a storm or there's a tripped breaker or another problem in an electrical circuit, understanding the basic components of an electrical system can help you get things running again. It's also important to know who is responsible for what portion of your electrical service. The utility company handles the line portion of your service, which includes everything up to the attachment point on your house. From there, it's called the load side, and everything on the load side is your responsibility.

Electrical Service Connection and Meter

Your home's electricity starts with the power service and electric meter. The utility company's service cables (whether overhead or underground) extend to your house and connect to the utility's electric meter. The meter measures the amount of electricity your home uses and is the basis for the charges on your electric bill. The meter runs only when electricity is used in the house.

Disconnect Switch

Many home electrical systems include a dedicated disconnect switch that is mounted on an outside wall of the home near the electric meter. In the event of a fire or flash flood, or if work needs to be done on the system, a disconnect switch allows you to shut off the power from outside the home so you don't have to enter the home to turn off the power. If an electrical system does not include a separate disconnect switch, the main circuit breaker in the home's main service panel (breaker box) serves as the system disconnect.

Main Service Panel

After passing through the meter, your electrical service feeds into your home's main service panel, commonly known as the breaker box. Two large "hot" wires connect to big screw terminals, called lugs, inside the service panel, providing all the power to the panel. A third service cable, the neutral, connects to the neutral bus bar inside the panel. In simple terms, electricity is supplied to the house on the hot wires. After it flows through the household system, it is fed back to the utility on the neutral wire, completing the electrical circuit.

Main Circuit Breaker

The service panel contains a large main breaker that is the switch controlling the power to the rest of the circuit breakers inside the panel. It is sized according to your home's service capacity. A standard panel today provides 200-amp (ampere) service. Older panels were sized for 150, 100, or fewer amps (amperes).

A main breaker of 200 amps will allow a maximum of 200 amps to flow through it without tripping. In a tripped state, no current will flow to the panel. In systems without an external disconnect switch, the main breaker serves as the household disconnect.

Turning off the main breaker stops the flow of power to all of the branch circuit breakers in the panel, and therefore to all of the circuits in the house. However, power is always flowing into the panel and to the service lugs even when the main breaker is shut off unless the power is shut off at a separate disconnect switch. Power is always present in the utility service lines and the electric meter unless it is shut off by the utility.

Branch Circuit Breakers

The breakers for the branch circuits fill the panel (usually below) the main breaker. Each of these breakers is a switch that controls the flow of electricity to a branch circuit in the house. Turning off a breaker shuts off the power to all of the devices and appliances on that circuit. If a circuit has a problem, such as an overload or a fault, the breaker automatically trips itself off.

The most common cause of a tripped breaker is a circuit overload. If you're running a high-demand appliance, such as a vacuum, toaster, or heater, and the power goes out, you've probably overloaded the circuit. Move the appliance to a different circuit and reset the breaker by switching it to the ON position. If the breaker trips again—without the appliance plugged in—you must call an electrician. There may be a dangerous fault situation in the circuit.

Devices

Devices are all the things in the house that use electricity, including switches, receptacles (outlets), light fixtures, and appliances. Devices are connected to the individual branch circuits that start at the breakers in the main service panel.

A single circuit may contain multiple switches, receptacles, fixtures, and other devices, or it may serve only a single appliance or receptacle. The latter is called a dedicated circuit. These are used for critical-use appliances, such as refrigerators, furnaces, and water heaters. Other appliances, such as dishwashers and microwaves, usually are on dedicated circuits, too, so that they can be shut off at the service panel without interrupting service to other devices. This also reduces the incidence of overloaded circuits.

Switches

Switches are the devices that turn on and off lights and fans in your home. They come in many different styles and colors to suit your design needs. There are single-pole, three-way, four-way, and dimmer switches. When you flip a switch off, it "opens" the circuit, meaning the circuit is broken or not complete and the power is interrupted. When the switch is on, the circuit is "closed," and power flows beyond the switch to the light or another device it is controlling.

Outlets

Electrical outlets, technically called receptacles, provide power to plug-in devices and appliances. Televisions, lights, computers, freezers, vacuums, and toasters are all good examples of devices that can be plugged into an outlet. Standard outlets in a home are either 15-amp or 20-amp; 20-amp outlets can provide more electricity without tripping a breaker. Special outlets for high-demand appliances, such as electric ranges and clothes dryers, may provide 30 to 50 or more amps of power.

In potentially wet areas of a home, such as bathrooms, kitchens, and laundry rooms, some or all of the outlets must have GFCI (ground-fault circuit-interrupter) protection, provided by GFCI outlets or a GFCI breaker.

Wiring

Your home's wiring consists of a few different types of wiring, including non-metallic cable (commonly called Romex), Bx cable, and wiring concealed in conduit. NM cable is the most common type of circuit wiring. It is suitable for use in dry, protected areas (inside stud walls, on the sides of joists, etc.) that are not subject to mechanical damage or excessive heat.

Bx cable, also known as armored cable, consists of wires running inside a flexible aluminum or steel sheath that is somewhat resistant to damage. It is commonly used where wiring for appliances, such as dishwashers and garbage disposers, is exposed.

Conduit is a rigid metal or plastic tubing that protects individual insulated wires. It is used in garages, sheds, and outdoor applications where the wiring must be protected from exposure.

Wires running inside NM cable, Bx cable, or conduit are sized according to each circuit's amperage. Wire size is given in its gauge number. The lower the gauge, the larger the wire, and the more current it can handle. For example, wiring for 20-amp circuits is 12-gauge, which is heavier than the 14-gauge wiring used for 15-amp circuits.

Indoor and Outdoor Electrical Wiring Safety Codes

Electrical codes are in place to protect you, the homeowner. These general guidelines apply to new installations and will give you the basics of what electrical inspectors are looking for. Be sure to check with your local electrical inspector because local codes may vary from the list provided. In the case of existing housing, the codes will apply if you are updating a home, and it requires an electrical update. It is also suggested that you update if the wiring in your home is unsafe and a danger to your family.

The National Electrical Code has some very specific rules and regulations about underground wiring methods and points of attachment. This is a look at the highlights of the outdoor sections of the code. Electrical wiring is often subjected to wet conditions and all of the elements that Mother Nature can throw at them. Electrical safety around swimming pools, hot tubs, and spas should be of extreme importance to the homeowner.

The NEC and Inspections

The National Electrical Code (NEC) was written to provide a set of rules and regulations to keep the use of electricity in your home safe. Here are the top bathrooms' codes you need to live by to remain safe and keep your electrical devices working properly.

You may wonder why the electrical inspector seems to be so tough on you when he explains all of the requirements in your bathroom. You may ask why you need things like GFCi's and exhaust fans. He may tell you that you need a separate circuit for your outlet, but after you consider everything that will probably be plugged into it, you soon see that the inspector is there to help you to have an effective and safe electrical plan.

Bathroom Electrical Codes

Each bathroom should have a circuit for lighting and an exhaust fan. This may include a blower-heater-light combination.

There should also be a 20-amp circuit, separate from the lighting circuit, to provide power for an outlet to feed things like curling irons, razors, hair dryers, and even portable milk house heaters.

Connected to the outlet circuit, you should install a ground fault circuit interrupter (GFCI) to protect the user. A GFCI trips and disconnects the circuit power if it senses a difference in potential on the circuit, like a short circuit or a path to ground, which could be right through your body. This device is very important and can save your life!

Since bathrooms are wet, switches should be grounded as well to give any stray voltage a direct path to ground, instead of through you.

You'd hate to get out of the shower, soaking wet, and get shocked by touching a switch.

Install at least one ceiling-mounted light fixture to allow ample lighting. This may be in addition to wall sconces or strip lighting in the bathroom.

Place exhaust fans or heater-fan-light combinations far enough from the bathtub, shower, or hot tub so that no one can stand in water and touch it.

Just remember, these are the bare minimum requirement, and you can add more circuits as you see fit to accommodate the load of the appliance you plan to plug in or add to your bathroom.

Kitchen

A kitchen should have a separate circuit for each appliance with a motor. The microwave, refrigerator, garbage disposal, and dishwasher would be the major appliances included. Generally, the code requires that you install a minimum of two receptacle circuits in the area above the countertop. An electric range, cooktop, or oven must be wired to a dedicated 240-volt circuit.

Living Room, Dining Room, and Bed Rooms

These rooms require that a wall switch is placed beside the entry door of the room so that you can light the room before entering it. It can either control a ceiling light, a wall light, or an outlet connected to a desk lamp. The ceiling fixture must be controlled by a wall switch and not a pull chain type light. Wall receptacles should be placed no farther than 12 feet apart. Dining rooms usually require a separate 20-amp circuit for one outlet used for a microwave, entertainment center, or window air conditioner.

Stairways

Special care is needed in stairways to ensure all of the steps are lighted properly. Three-way switches are required at the top and bottom of the stairs. If the stairs turn, you may need to add additional lighting to accommodate the area to be lit.

Hallways

These areas can be long and need adequate lighting. Be sure to place enough lighting so shadows are not cast when walking. Remember, hallways are often escape routes in the event of inclement weather and emergencies. A hallway over 10 feet long is required to have an outlet for general purpose. Three-way switches are required for the two ends of the hallway. If there are more doors throughout the hallway, say a bedroom or two, then you may want to add addition four-way switches to the circuit outside the door of each room.

Closets

Closets must have one globe covered fixture controlled by a wall switch. Exposed bulb fixtures, like pull-chain fixtures, get hot and come in contact with clothing or other combustible materials stored in closets. Although your existing home may have these fixtures, it is recommended that you change them for safety reasons.

Laundry Room

The washer and dryer should have its own 20-amp receptacle. In the case of an electric dryer, a separate 240-volt circuit should be installed.

Attached Garage

Inside the garage, there should be at least one switch controlling the lighting. It is recommended that three-way switches be installed for convenience between the doors. This lighting should be in addition to any garage door lighting that you may have. Garages need a separate circuit for at least one outlet. This is generally required to be a GFCI outlet. You should check your local code to be sure. When in doubt, make it a GFCI.

Any outside outlets connected must be either a GFCI outlet or an outlet connected to a GFCI breaker.

Remember that the electrical codes are in place for your safety. Although you may believe that they are overkill at times, these practices save lives every day. When it comes to electrical safety, don't become a statistic! Follow the rules of the codes and be sure to have your local electrical inspector give you the green light for the safety of your family's sake.

Checking for Incorrect Electrical Wiring

By identifying electrical wiring hazards before problems appear, you can make your home safer and possibly prevent a fire or a dangerous electrical shock. Even the humble electrical outlet or light switch can have numerous things that can go wrong, most of them resulting from faulty installation. Here, then, is a list of wiring problems you might encounter by simply peering into an outlet or switch box with a flashlight. Many of these are easy to fix, but if you find a lot of them, you might want to call in an electrician for an expert inspection of your entire electrical system.

Safety First... Turn off the Power

Before working on any electrical circuit or device, always turn off the power to the entire circuit by switching off the appropriate breaker in your home's service panel (breaker box).

After you've switched off the breaker, test any circuit wires or devices you'll be inspecting with a non-contact voltage tester. This inexpensive tool is about the size and shape of a permanent marker and allows you to test for power without touching any wires. Simply touch the tip of the tester to the wire in question (or insert the tip into an outlet slot or touch it to any device terminal). The tester can detect voltage through the wiring insulation, so you don't have to find the bare end of the wire, as you do with some other testers. If there's voltage, the tester lights up. No light, no voltage.

Reversed Connections

Most electrical outlets (properly called receptacles) today are grounded three-prong outlets. They have one long straight slot, one short straight slot and a roundish ground slot to accept the three prongs of a grounded plug. Older, ungrounded, outlets have only two straight slots, one long and one short.

That's why you often have to flip over a plug to fit it into an outlet; it goes in only one way. This long/short design is called polarized and is a safety feature that predates the standard grounded outlet.

Polarized outlets and plugs ensure that electricity flows in one direction only. This makes things like lamps and many appliances more safe to operate. But here's the catch: If you connect the circuit wires to the wrong terminals on an outlet, the outlet will still work but the polarity will be backward. When this happens, a lamp, for example, will have its bulb socket sleeve energized rather than the little tab inside the socket. Guess which you're more likely to touch? You want the tab energized, not the sleeve.

Inside an outlet's electrical box, the black (hot) wire should be connected to the brass-colored terminal on the outlet. The white (neutral) wire should be connected to the silver-colored terminal. If these connections are backward, the polarity is wrong.

Proper Grounding

In a modern home, almost every part of the electrical system is grounded, meaning it has an unbroken (if usually not direct) connection to the earth outside the house. When something goes wrong, such as a short or fault, electricity flows safely to the earth via the grounding system.

Homes that date back to the 1950s and earlier may have few or no true ground connections. Is this dangerous? It can be. Sometimes very dangerous. But the fact is, most of these homes operate just fine without grounded circuits. That said, if you're adding new circuits or updating any part of an electrical system, you should always include a ground. It's not just smart; it's the law.

If you have outlets on ungrounded circuits you can replace them with GFCIs, or ground-fault circuit-interrupters. These are special outlets that shut off the power if they detect a dangerous ground fault, helping to protect you against shock. They do NOT provide a ground, but they do make using the outlet a lot safer.

One simple way to test outlets for grounding is to plug in a receptacle tester. If the tester indicates an "open ground," the outlet may have no means of grounding or there may be a ground wire but it's improperly connected. It also could be grounded to a metal electrical box but the box is not properly grounded.

Too Many Wires Under Terminals

Installing more than one wire under any standard screw terminal is not only a stupid move, it's a lazy one at that. It is nearly impossible to properly tighten two wires under a single terminal. This usually results in a loose connection. And loose wires are a very bad thing. If you find more than one wire connected to an outlet or switch, correct the problem by joining the wires with a wire connector and include a pigtail, a short length of the same type of wire. Connect the pigtail-- and only the pigtail--to the screw terminal in ⁨uestion.

Proper Amount of Wire Insulation

Although it may not seem important, proper insulation length is very important on wire connection points. Stripping a wire to the proper length makes for a great connection. Stripping too much insulation exposes the bare wire too much and can become a point where someone can touch the wire or the bare wire may come in contact with the box or another wire, like the ground wire. in this case, some folks just cover the exposed wires with electrical tape, but the proper method is to re-strip the wire end to proper length.

Too little wire shouldn't be a problem then, right? Wrong! Too little insulation means that some or all of the terminal is in contact with the insulation and not the bare wire. This either means that there is a limited connection, with resistance due to the insulation, or no connection at all.

When stripping wire for a screw terminal, remove about 3/4 inch of insulation from the wire end. Shape the bare wire end into a hook and attach it to the terminal so the open end of the hook is on the right; this means the hook tightens around the screw as the screw is turned. When the connection is complete, the wire insulation should almost touch the screw, but none of it should be under the screw.

Common Wire Connection Problems and Their Solutions

A great many electrical problems around the house are traced to different versions of the same essential problem: wire connections that are made improperly or that have loosened over time. You may have inherited the problem from a previous owner or from an electrician who did an inadequate job, or perhaps it's the result of work you did yourself. Many wire connection problems are no one's fault but are simply the result of time. Wires are under a constant cycle of heating and cooling, expansion and contraction. Every time a switch is used or appliances are plugged in, and the natural result of all this usage is that wire connections can loosen over time.

Your electrical system has a lot of safeguards against danger from bad wire connections, such as its grounding system, its circuit breakers, and GFCI and AFCI protection. Still, there is danger from sparking and arcing whenever there is a loose wire connection in your system. Many of these problems are quite easy for a homeowner to spot and repair, while others are best handled by a professional electrician. Understanding where these problems commonly occur will help you decide how to handle them.

Tools and Materials
• Flashlight
• Wire strippers
• Screwdrivers
• Utility knife
• Wire connectors (wire nuts)
• Eye protection
• Electrical wire in various gauges

Here are six very common places that wire connection problems occur.

Loose Wire Connections at Switches and Outlets

By far the most common problem is when screw terminal connections at wall switches and outlets become loose. Because these fixtures get the most use within an electrical system, these are the places to look first if you suspect wire connection problems.

Loose wire connections at a switch, outlet, or light fixture are often signaled by a buzzing or crackling sound or by a light fixture that flickers.

To address this problem, it involves first turning off the power to the suspected wall switch, light fixture, or outlet. With the power shut off, you can remove the cover plate and use a flashlight to carefully examine the screw terminals inside where the wires are connected. If you find any that are loose, carefully tighten the screw terminals down onto the wires. In all likelihood, this will fix the problem.

Sometimes, you may find that the wire connections are made via push-in fittings on the back of the switch or outlet. This method of connection is notorious for being prone to failure—so much so that most professional electricians don't use the push-in fittings at all, but instead make all wire connections with the screw terminal connections on the sides of the switch or outlet. If you find that your device is made with the push-in fittings, you might want to remove them and reconnect the wires to the screw terminals on the device.

Finally, if there are pass-through wire connections inside the box that are made with wire nuts or another type of connector, check these to make sure the wires are tightly joined together. A loose connector is also a common source of problems.

Wire Connections Made With Electrical Tape

A classic wire connection error is when wires are joined together with electrical tape rather than a wire nut or other sanctioned connector.

To fix the problem, first, turn off the power to the circuit. Then, remove the electrical tape from the wires and clean them. Make sure there is the proper amount of exposed wire showing (for most connectors, this means about 3/4 inch), then join the wires together with a wire nut or other approved connector (there are now push-in connectors that some pros like to use).

If the wire ends are damaged, you can cut off the ends of the wires and strip off about 3/4 inch of insulation to make a proper wire nut connection.

Two or More Wires Under One Screw Terminal

Another common wire connection problem is when you find two or more wires held under a single screw terminal on a switch or outlet. This is a clear sign of amateur work and a distinct fire hazard. It is allowable to have a single wire under each of the two screw terminals on the side of an outlet or switch, but it is a code violation to have two wires wedged under a single screw. This is most often seen when two bare copper grounding wires are found under the grounding screw on the outlet or switch, but you also may occasionally find hot wires or neutral wires connected to a single screw terminal.

To fix this problem, once again, this repair involves first shutting off the power. Then, the two offending wires are removed from their screw terminal. Cut a 6-inch pigtail wire of the same color as the two wires (use a green pigtail if you are joining two bare copper grounding wires). Strip 3/4 inch of insulation from each end of the pigtail, then join one end to the two wires you just disconnected, using a wire connector (wire nut). Now, attach the free end of the pigtail wire to the screw terminal that once held the two wires.

You have essentially created a bridge, or pathway, that connects both wires to the desired screw terminal on the outlet or switch.

Note: Make sure the pigtail wire is the same wire gauge as the circuit wires. A 15-amp circuit normally uses 14-gauge wire; a 20-amp circuit uses 12-gauge wire.

Exposed Wires

It is quite common, especially with amateur electrical work, to see a screw terminal connection or wire nut connection where it has too much (or too little) exposed copper wire showing at the wires. With screw terminal connections, there should be enough bare copper wire stripped to wrap entirely around the screw terminal but not so much that excess bare copper wire extends out from the screw. The excess exposed wire can short out if it touches a metal box or other wires. Wires should be wrapped clockwise around the screw terminals; if they are reversed, they can be prone to loosening.

With wire nut connections, all of the bare copper wire should be hidden under the plastic cap, with no exposed wire showing at the bottom of the wire nut.

To fix the problem, turn off the power to the device, then disconnect the wires and either clip off the excess wire or strip off additional insulation so the proper amount of wire is exposed. Then, reconnect the wires to their screw terminal or wire nut. Tug lightly on the wires to make sure they are securely connected.

Loose Connections on Circuit Breaker Terminals

A less common problem is when the hot wires on circuit breakers in the main service panel are not tightly connected to the breaker. When this happens, you may notice lights flickering or service problems on fixtures all along the circuit. When making connections to circuit breakers, be sure to strip the proper amount of wire insulation from the wire and make sure that only the bare wire is placed under the terminal slot before tightening. Insulation under the connection slot is a code violation.

To fix the problem, repairs at the main service panel should be handled by a professional electrician. Amateurs should attempt these repairs only if they are ⬚uite experienced and knowledgeable about electrical systems.

The electrician will address this problem by turning off the breaker then unclipping it from the hot bus bar in the main service panel. He or she will check the hot wire connected to the breaker to make sure that the screw is tight and that there is no insulation under the terminal and no excess bare copper wire exposed. With repair complete, the electrician will snap the breaker back into place on the hot bus bar and turn the breaker back on.

Faulty Neutral Wire Connections at Circuit Breaker Panels

Another less common problem—and another that is usually handled by a pro—is when the white circuit wire is not correctly mounted to the neutral bus bar in the main service panel. Symptoms here will be similar to those with a faulty hot wire.

To fix this problem, the electrician will check to make sure the neutral wire is sufficiently stripped and correctly attached to the neutral bus bar.

Chapter 8: What Is Electrical Wiring?

Electrical wiring is an electrical installation of cabling and associated devices such as switches, distribution boards, sockets, and light fittings in a structure. Wiring is subject to safety standards for design and installation. Allowable wire and cable types and sizes are specified according to the circuit operating voltage and electric current capability, with further restrictions on the environmental conditions, such as ambient temperature range, moisture levels, and exposure to sunlight and chemicals.

Associated circuit protection, control and distribution devices within a building's wiring system are subject to voltage, current and functional specification. Wiring safety codes vary by locality, country or region. The International Electro-Technical Commission (IEC) is attempting to harmonize wiring standards amongst member countries, but significant variations in design and installation re[?]uirements still exist.

Electrical Wiring is a process of connecting cables and wires to the related devices such as fuse, switches, sockets, lights, fans etc. To the main distribution board is a specific structure to the utility pole for continues power supply.

Early wiring methods

The first interior power wiring systems used conductors that were bare or covered with cloth, which were secured by staples to the framing of the building or on running boards. Where conductors went through walls, they were protected with cloth tape. Splices were done similarly to telegraph connections, and soldered for security. Underground conductors were insulated with wrappings of cloth tape soaked in pitch, and laid in wooden troughs which were then buried. Such wiring systems were unsatisfactory because of the danger of electrocution and fire, plus the high labor cost for such installations. The first Electrical codes arose in the 1880s with the commercial introduction of electrical power; however, many conflicting standards existed for the selection of wire sizes and other design rules for electrical installations, and a need was seen to introduce uniformity on the grounds of safety.

The earliest standardized method of wiring in buildings, in common use in North America from about 1880 to the 1930s, was knob and tube (K&T) wiring: single conductors were run through cavities between the structural members in walls and ceilings, with ceramic tubes forming protective channels through joists and ceramic knobs attached to the structural members to provide air between the wire and the lumber and to support the wires. Since air was free to circulate over the wires, smaller conductors could be used than re□uired in cables. By arranging wires on opposite sides of building structural members, some protection was afforded against short-circuits that can be caused by driving a nail into both conductors simultaneously.

By the 1940s, the labor cost of installing two conductors rather than one cable resulted in a decline in new knob-and-tube installations. However, the US code still allows new K&T wiring installations in special situations (some rural and industrial applications).

In the United Kingdom, an early form of insulated cable, introduced in 1896, consisted of two impregnated-paper-insulated conductors in an overall lead sheath. Joints were soldered, and special fittings were used for lamp holders and switches. These cables were similar to underground telegraph and telephone cables of the time. Paper-insulated cables proved unsuitable for interior wiring installations because very careful workmanship was required on the lead sheaths to ensure moisture did not affect the insulation.

A system later invented in the UK in 1908 employed vulcanized-rubber insulated wire enclosed in a strip metal sheath. The metal sheath was bonded to each metal wiring device to ensure earthing continuity. A system developed in Germany called "Kuhlo wire" used one, two, or three rubber-insulated wires in a brass or lead-coated iron sheet tube, with a crimped seam. The enclosure could also be used as a return conductor. Kuhlo wire could be run exposed on surfaces and painted, or embedded in plaster. Special outlet and junction boxes were made for lamps and switches, made either of porcelain or sheet steel. The crimped seam was not considered as watertight as the Stannos wire used in England, which had a soldered sheath.

A somewhat similar system called "concentric wiring" was introduced in the United States around 1905. In this system, an insulated electrical wire was wrapped with copper tape which was then soldered, forming the grounded (return) conductor of the wiring system. The bare metal sheath, at earth potential, was considered safe to touch. While companies such as General Electric manufactured fittings for the system and a few buildings were wired with it, it was never adopted into the US National Electrical Code. Drawbacks of the system were that special fittings were re🗆uired, and that any defect in the connection of the sheath would result in the sheath becoming energized.

Armored cables with two rubber-insulated conductors in a flexible metal sheath were used as early as 1906, and were considered at the time a better method than open knob-

and-tube wiring, although much more expensive.

The first rubber-insulated cables for US building wiring were introduced in 1922 with US patent 1458803, Burley, Harry & Rooney, Henry, "Insulated electric wire", issued 1923-06-12, assigned to Boston Insulated Wire and Cable. These were two or more solid copper electrical wires with rubber insulation, plus woven cotton cloth over each conductor for protection of the insulation, with an overall woven jacket, usually impregnated with tar as a protection from moisture. Waxed paper was used as a filler and separator.

Over time, rubber-insulated cables become brittle because of exposure to atmospheric oxygen, so they must be handled with care and are usually replaced during renovations. When switches, socket outlets or light fixtures are replaced, the mere act of tightening connections may cause hardened insulation to flake off the conductors. Rubber insulation further inside the cable often is in better condition than the insulation exposed at connections, due to reduced exposure to oxygen.

The sulfur in vulcanized rubber insulation attacked bare copper wire so the conductors were tinned to prevent this. The conductors reverted to being bare when rubber ceased to be used.

About 1950, PVC insulation and jackets were introduced, especially for residential wiring. About the same time, single conductors with a thinner PVC insulation and a thin nylon jacket (e.g. US Type THN, THHN, etc.) Became common.

The simplest form of cable has two insulated conductors twisted together to form a unit. Such non-jacketed cables with two (or more) conductors are used only for extra-low voltage signal and control applications such as doorbell wiring.

Other methods of securing wiring that are now obsolete include:
Re-use of existing gas pipes when converting gas lighting installations to electric lighting. Insulated conductors were pulled through the pipes that had formerly supplied the gas lamps. Although used occasionally, this

method risked insulation damage from sharp edges inside the pipe at each joint.

Wood mouldings with grooves cut for single conductor wires, covered by a wooden cap strip. These were prohibited in North American electrical codes by 1928. Wooden mouldings was also used to some degree in the UK, but was never permitted by German and Austrian rules.

A system of flexible twin cords supported by glass or porcelain buttons was used near the turn of the 20th century in Europe, but was soon replaced by other methods.

During the first years of the 20th century, various patented forms of wiring system such as Bergman and Peschel tubing were used to protect wiring; these used very thin fibre tubes, or metal tubes which were also used as return conductors.

In Austria, wires were concealed by embedding a rubber tube in a groove in the wall, plastering over it, then removing the tube and pulling wires through the cavity.

Metal moulding systems, with a flattened oval section consisting of a base strip and a snap-on cap channel, were costlier than open wiring or wooden moulding, but could be easily run on wall surfaces. Similar surface mounted raceway wiring systems are still available today.

Methods of Electrical Wiring Systems w.r.t Taking Connection

Wiring (a process of connecting various accessories for distribution of electrical energy from supplier's meter board to home appliances such as lamps, fans and other domestic appliances is known as Electrical Wiring) can be done using two methods which are:

Joint box system or Tee system

Loop – in system

They are discussed as follows:
Joint Box or Tee or Jointing System

In this method of wiring, connections to appliances are made through joints. These joints are made in joint boxes by means of suitable connectors or joints cutouts. This method of wiring doesn't consume too much cables size.

You might think because this method of wiring doesn't re🔲uire too much cable it is therefore cheaper. It is of course but the money you saved from buying cables will be used in buying joint boxes, thus e🔲uation is balanced. This method is suitable for temporary installations and it is cheap.

Loop-in or Looping System
This method of wiring is universally used in wiring. Lamps and other appliances are connected in parallel so that each of the appliances can be controlled individually. When a connection is required at a light or switch, the feed conductor is looped in by bringing it directly to the terminal and then carrying it forward again to the next point to be fed.

The switch and light feeds are carried round the circuit in a series of loops from one point to another until the last on the circuit is reached. The phase or line conductors are looped either in switchboard or box and neutrals are looped either in switchboard or from light or fan. Line or phase should never be looped from light or fan.

Advantages of Loop-In Method of Wiring
1. It doesn't require joint boxes and so money is saved

2. In loop – in systems, no joint is concealed beneath floors or in roof spaces.

3. Fault location is made easy as the points are made only at outlets so that they are accessible.

Disadvantages of Loop-In Method of Wiring
1. Length of wire or cables required is more and voltage drop and copper losses are therefore more

2. Looping – in switches and lamp holders is usually difficult.

Different Types of Electrical Wiring Systems
The types of internal wiring usually used are:

- Cleat wiring

- Wooden casing and capping wiring

- CTS or TRS or PVC sheath wiring

- Lead sheathed or metal sheathed wiring
- Conduit wiring

There are additional types of conduit wiring according to Pipes installation (Where steel and PVC pipes are used for wiring connection and installation).

- Surface or open Conduit type

- Recessed or concealed or underground type Conduit

Cleat Wiring

This system of wiring comprises of ordinary VIR or PVC insulated wires (occasionally, sheathed and weather proof cable) braided and compounded held on walls or ceilings by means of porcelain cleats, Plastic or wood. Cleat wiring system is a temporary wiring system therefore it is not suitable for domestic premises. The use of cleat wiring system is over nowadays.

In this, porcelain, wood or plastic cleats are fixed to walls or ceilings at regular intervals, i.e., 0.6 m between each cleat. PVC insulated cables are taken through the holes of each cleat and hence cleat support and holds wire.

This is an inexpensive method of wiring and is used for temporary installations. Therefore, it is not suitable for home electrical wiring and also it is an outdated method.

Advantages of Cleat Wiring:
1. It is simple and cheap wiring system

2. Most suitable for temporary use i.e. Under construction building or army camping

3. As the cables and wires of cleat wiring system is in open air, therefore fault in cables can be seen and repair easily.

4. Cleat wiring system installation is easy and simple.

5. Customization can be easily done in this wiring system e.g. Alteration and addition.

6. Inspection is easy and simple.

Disadvantages of Cleat Wiring:
1. Appearance is not so good.

2. Cleat wiring can't be use for permanent use because, Sag may be occurring after sometime of the usage.

3. In this wiring system, the cables and wiring is in open air, therefore, oil, Steam, humidity, smoke, rain, chemical and acidic effect may damage the cables and wires.

4. It is not lasting wire system because of the weather effect, risk of fire and wear & tear.

5. It can be only used on 250/440 Volts on low temperature.

6. There is always a risk of fire and electric shock.

7. It can't be used in important and sensitive location and places.

8. It is not lasting, reliable and sustainable wiring system.

Casing and Capping wiring
Casing and Capping wiring system was famous wiring system in the past but, it is considered obsolete this days because of Conduit and sheathed wiring system. The cables used in this kind of wiring were either VIR or PVC or any other approved insulated cables.

The cables were carried through the wooden casing enclosures. The casing is made up of a strip of wood with parallel grooves cut length wise so as to accommodate VIR cables. The grooves were made to separate opposite polarity. The capping (also made of wood) used to cover the wires and cables installed and fitted in the casing.

In this cable is run through a wood casing having grooves. The wood casing is prepared in such a way that it is of a required fixed length with parallel grooves that accommodates the cables. The wooden casing is fixed to the walls or ceiling with screws.

After placing the cables inside the grooves of casing, a wooden cap with grooves is placed on it to cover the cables. This is also a cheap wiring system, but there is a high risk of fire in case of short circuits.

Advantages of Casing Capping Wiring:
1. It is cheap wiring system as compared to sheathed and conduit wiring systems.

2. It is strong and long-lasting wiring system.

3. Customization can be easily done in this wiring system.

4. If Phase and Neutral wire is installed in separate slots, then repairing is easy.

5. Stay for long time in the field due to strong insulation of capping and casing..

6. It stays safe from oil, Steam, smoke and rain.

7. No risk of electric shock due to covered wires and cables in casing & capping.

Disadvantages Casing Capping Wiring:
1. There is a high risk of fire in casing & capping wiring system.

2. Not suitable in the acidic, alkalies and humidity conditions

3. Costly repairing and need more material.

4. Material can't be found easily in the contemporary

5. White ants may damage the casing & capping of wood.

Batten Wiring (CTS or TRS)
Single core or double core or three core TRS cables with a circular oval shape cables are used in this kind of wiring. Mostly, single core cables are preferred. TRS cables are chemical proof, water proof, steam proof, but are slightly affected by lubricating oil. The TRS cables are run on well seasoned and straight

teak wood batten with at least a thickness of 10mm.

The cables are held on the wooden batten by means of tinned brass link clips (buckle clip) already fixed on the batten with brass pins and spaced at an interval of 10cm for horizontal runs and 15cm for vertical runs.

In this, insulated wires are run through the straight teak wooden battens. The wooden battens are fixed on the ceilings or walls by plugs and screws. The cables are fitted onto the battens by using tinned brass link clips.

These clips are fixed to the battens with rust-resistant nails. This wiring installation is simple and cheap as compared to other electrical wiring systems also takes less time to install. These are mainly used for indoor installations.

Advantages of Batten Wiring
1. Wiring installation is simple and easy

2. Cheap as compared to other electrical wiring systems

3. Paraphrase is good and beautiful

4. Repairing is easy

5. Strong and long-lasting

6. Customization can be easily done in this wiring system.

7. Less chance of leakage current in batten wiring system

Disadvantages of Batten Wiring
1. Can't be install in the humidity, Chemical effects, open and outdoor areas.

2. High risk of firs

3. Not safe from external wear & tear and weather effects (because, the wires are openly visible to heat, dust, steam and smoke.

4. Heavy wires can't be used in batten wiring system.

5. Only suitable below then 250V.

6. Need more cables and wires.

Lead Sheathed Wiring

The type of wiring employs conductors that are insulated with VIR and covered with an outer sheath of lead aluminum alloy containing about 95% of lead. The metal sheath given protection to cables from mechanical damage, moisture and atmospheric corrosion.

The whole lead covering is made electrically continuous and is connected to earth at the point of entry to protect against electrolytic action due to leaking current and to provide safety in case the sheath becomes alive. The cables are run on wooden batten and fixed by means of link clips just as in TRS wiring.

Conduit Wiring

In this wiring, PVC cables are taken through either PVC conduit pipes or through steel conduit pipes. This conduit wiring can be either surface conduit wiring or concealed conduit wiring.

If the conduit pipes are run on surface of the walls and ceilings, it is called a surface conduit wiring. If the conduits are run inside the surface of the walls and ceilings and are covered with plastering, it is called as concealed conduit wiring.

There are two additional types of conduit wiring according to pipe installation

• Surface Conduit Wiring

• Concealed Conduit Wiring

Surface Conduit Wiring

If conduits installed on roof or wall, it is known as surface conduit wiring. In this wiring method, they make holes on the surface of wall on eᵭual distances and conduit is installed then with the help of real plugs.

Concealed Conduit wiring
If the conduits are hidden inside the wall slots with the help of plastering, it is called concealed conduit wiring. In other words, the electrical wiring system inside wall, roof or floor with the help of plastic or metallic piping

is called concealed conduit wiring. Obliviously, it is the most popular, beautiful, stronger and common electrical wiring system nowadays.

Conduit Steel Non-metallic Conduit PVC
In conduit wiring, steel tubes known as conduits are installed on the surface of walls by means of pipe hooks (surface conduit wiring) or buried in walls under plaster and VIR or PVC cables are afterwards drawn by means of a GI wire of size if about 18SWG.

In Conduit wiring system, the conduits should be electrically continuous and connected to earth at some suitable points in case of steel conduit. Conduit wiring is a professional way of wiring a building. Mostly PVC conduits are used in domestic wiring.

The conduit protects the cables from being damaged by rodents (when rodents bites the cables it will cause short circuit) that is why circuit breakers are in place though but hey! Prevention is better than cure. Lead conduits are used in factories or when the building is prone to fire accident. Trunking is more of like surface conduit wiring. It's gaining popularity too.

It is done by screwing a PVC trunking pipe to a wall then passing the cables through the pipe. The cables in conduit should not be too tight. Space factor have to be put into consideration.

Types of Conduit
Following conduits are used in the conduit wiring systems (both concealed and surface conduit wiring).

Metallic Conduit

Non-metallic conduit

Metallic Conduit:

Metallic conduits are made of steel which are very strong but costly as well.

There are two types of metallic conduits:
Class A Conduit: Low gauge conduit (Thin layer steel sheet conduit)

Class B Conduit: High gauge conduit (Thick sheet of steel conduit)

Non-metallic Conduit:

A solid PVC conduit is used as non-metallic conduit nowadays, which is flexible and easy to bend.

Size of Conduit:

The common conduit pipes are available in different sizes genially, 13, 16.2, 18.75, 20, 25, 37, 50, and 63 mm (diameter) or 1/2, 5/8, 3/4, 1, 1.25, 1.5, and 2 inch in diameter.

Advantage of Conduit Wiring Systems
1. It is the safest wiring system (Concealed conduit wring)

2. Appearance is very beautiful (in case of concealed conduit wiring)

3. No risk of mechanical wear & tear and fire in case of metallic pipes.

4. Customization can be easily done according to the future needs.

5. Repairing and maintenance is easy.

6. There is no risk of damage the cables insulation.

7. It is safe from corrosion (in case of PVC conduit) and risk of fire.

8. It can be used even in humidity , chemical effect and smoky areas.

9. No risk of electric shock (In case of proper earthing and grounding of metallic pipes).

10. It is reliable and popular wiring system.

11. Sustainable and long-lasting wiring system.

Disadvantages of Conduit Wiring Systems
1. It is expensive wiring system (Due to PVC and Metallic pipes, Additional earthing for metallic pipes Tee(s) and elbows etc.

2. Very hard to find the defects in the wiring.

3. Installation is not easy and simple.

4. Risk of electric shock (In case of metallic pipes without proper earthing & grounding system)

5. Very complicated to manage additional connection in the future.

Electrical Safety
Before starting any installation work, first and foremost thing is the concern of safety of the personnel. Electricity is dangerous, direct or indirect contact of electrical equipment or wires with the power turned ON can result serious injuries or sometimes even causes to death. Follow the below steps to maintain the safety at the workplace.

Guides to follow:
1. Always use safety equipment like goggles, gloves, shoes, etc. And avoid the direct contact with live or energized circuits.

2. Have the skills and techniques to distinguish the exposed live parts of the electrical equipment.

3. Disconnect the source supply while installing or connecting wires.

4. The power supplied to the installation must be controlled on the main switchboard which should consist of circuit breaker.

5. Conductive tools and materials must be kept at a safe distance from live parts of the circuit or equipment.

6. Use non-conductive hand tools for which they are rated to perform electrical work. If they are used for voltage (or current) rating other than rated, the insulation strength of the tool breakdown and causes electric shock.

Conclusion

Sometimes in life when we get down to crunch time—right down to the wire—we develop a singular focus and can't put our energy in much else. But as this book has demonstrated, home wiring does not have to be a struggle. The world of the DIY electrician is not limited, and there are plenty of options available for you.

All you really need is a little patience and perseverance and all the rest will fall into place for you in short order. Even if it seems tough to get started on big wiring projects— just like anything else—if you give it enough time, and trial and error, you will be able to master it. So, take the time to learn this valuable skill. You may be down to the wire, but even so, that doesn't mean you can't install some fabulous wiring into your home. Thank you for reading!